RESEARCH IN MARITIME HISTORY
NO. 53

PLANNING AND PROFITS

BRITISH NAVAL ARMAMENTS MANUFACTURE AND THE MILITARY-INDUSTRIAL COMPLEX, 1918–1941

CHRISTOPHER W. MILLER

MAILING ADDRESS

Liverpool University Press
4 Cambridge Street
Liverpool L69 7ZU
United Kingdom

RESEARCH IN MARITIME HISTORY
NO. 53

Planning and Profits

British Naval Armaments Manufacture and the Military-Industrial Complex, 1918–1941

CHRISTOPHER W. MILLER

First published 2018 by
Liverpool University Press
4 Cambridge Street
Liverpool
L69 7ZU

British Library Cataloguing-in-Publication data
A British Library CIP record is available

ISBN 978-1-78694-066-7

Typeset by Carnegie Book Production, Lancaster
Printed and bound by TJ International Ltd, Padstow, Cornwall, PL28 8RW

Contents

Part Four: The Ultimate Potential Enemy and Rearmament Planning, 1934–1936

Part Five: The Inskip Era and War, 1936–1941

Acknowledgements

This book began life as a doctoral thesis at the University of Glasgow, though it also owes something to the ideas in a master's dissertation on rearmament at the University of Oxford. At Glasgow, I am principally indebted to Professor Raymond Stokes, director of the Centre for Business History in Scotland, and Professor Phillips O'Brien, onetime director of the Scottish Centre for War Studies, and now at the University of St. Andrews. Dr. Nicholas Rodger, of All Souls College, Oxford, guided my earliest forays into the shipbuilding collections at the National Maritime Museum, while Professor Simon Ball (now at Leeds University) did the same for the Committee of Imperial Defence collections in the National Archives. Thanks are also due to my examiners, Professor Peter Jackson (Glasgow) and Professor George Peden (Stirling), whose thoroughness and supportive comments further convinced me of the project's potential. I consider myself extremely fortunate to have been assisted by such first-rate minds; without them I am fully convinced that this work would never have made it this far.

For turning this research into the book before you today, I am principally indebted to my editors Professor Lewis (Skip) Fischer and Professor Hugh Murphy, who saw through the messier aspects of the thesis to the underlying promise and a finished manuscript. They, along with Alison Welsby at Liverpool University Press, have remained enormously patient and helpful while numerous versions and unfinished work were batted back and forward to them, never failing to improve upon my prose each time.

This research could not have been undertaken had it not been for a full scholarship from the Arts and Humanities Research Council, for which I remain deeply appreciative. In addition, the Economic History Society, the Lind Foundation, the Snell Exhibition, the University of Glasgow's College of Arts, and Yale University, have supported this work materially and financially. In the last case, the Department of International Security Studies and Professor Paul Kennedy were gracious and supportive hosts during a fellowship in 2012. My former employers at the University of Tübingen, and Dr. Daniel Menning in particular, provided an intellectually stimulating environment in which to turn the thesis into the substantially revised book before you now.

I have been assisted from the outset by a group of very helpful – and patient – archivists and librarians. Those at the Scottish Business Archives at the University of Glasgow, the Mitchell Library, the National Archives (Kew),

Churchill College Cambridge and the University of Cambridge Library, and at the Caird Library of the National Maritime Museum (Greenwich) have made the arduous task of sifting through mountains of material much easier to bear.

Finally, the greatest debt of gratitude is to my family, and especially to my partner Enni. In practical terms, she has read chapters, drafts, fixed typographical errors, and endured my extended soliloquys with characteristic good humour. More fundamentally, she has been a constant source of inspiration and encouragement throughout the past seven years. Without her, the task would have been immeasurably harder.

All errors in this work are entirely my own responsibility.

Glasgow, 2017

About the Author

Dr. Christopher W. Miller was educated at Balliol College, Oxford University and The University of Glasgow, where he currently works at the Centre for Business History in Scotland. Prior to returning to Scotland, he was lecturer in Modern History at the University of Tübingen. He has published on a variety of business and labour history topics, including imperialism, cartels, elite networks, radical politics, and pollution from the 1880s to the 1960s. This is his first book.

Tables and Figures

Illustrations

between pages 121 and 122

All images reproduced with the kind permission of the Senior Archivist, Glasgow University Archives Service.

Abbreviations

ADM: Admiralty papers (The National Archives, Kew, London)
BIDC: Bankers' Industrial Development Company
BLH: Basil Liddell Hart Centre for Military Archives, King's College London
BOE: Bank of England
CAB: Cabinet Papers (The National Archives)
CID: Committee of Imperial Defence
CCC: Churchill College Archives Centre, University of Cambridge
CoS: Chiefs of Staff Subcommittee
CUL: Cambridge University Library, Manuscripts Room
DBFP: Documents on British Foreign Policy
DNC: Director of Naval Construction
DPR: Defence and Policy Requirements Subcommittee
DRC: Defence Requirements Committee
GCC: Glasgow City Council Archives (Mitchell Library)
GUAS: Glasgow University Archives (includes the Scottish Business Archive)
MFCD: Minister for the Coordination of Defence
NMM: National Maritime Museum, Greenwich
NSS: National Shipbuilders' Security
PSOC: Principal Supply Officers Committee
SCI, II, III etc.: Supply Committee I, II, III
SEF: Shipbuilding Employers' Federation
SMT: Securities Management Trust
SUPP: Ministry of Supply Papers (in the National Archives)
TNA: The National Archives, Kew, London
WSBC: Warshipbuilders' Committee

A Note on Definitions

Although a great many firms made support vessels for the Royal Navy, some definition of "warships" is useful in drawing boundaries between the private naval armaments industry and smaller vessels typically made in non-specialist shipbuilding yards. Where grey areas occur, they mainly concern ships smaller than destroyers (i.e., under about 1000 tons), which were most commonly sloops-of-war, minesweepers, frigates and corvettes. Confusingly, the Royal Navy frequently dropped and resurrected terminology, leading some classes such as the *Black Swan* to be popularly known as "sloops" while officially designated "frigates."[1] Fortunately, the years between the wars are largely free from most of these problems, as it was not until the final months before the Second World War broke out that mercantile shipbuilders took over the task of building very small escort and support ships to free up space for more demanding tasks elsewhere. Until that point it was the recognised warship manufacturers – those that feature in this book – that undertook such orders, constructing even the smaller ships to a high specification unsuitable for mass production. The same can also be said for the various submarine classes of the period.[2] Therefore, when general statements about warships and naval shipbuilding are made hereafter they can be safely applied to almost every ship fitted with weaponry launched for the Royal Navy during these years.

---

[1] A. Hague, *Sloops: A History of the 71 Sloops Built in Britain and Australia for the British, Australian and Indian and Indian Navies, 1926–1946* (Kendal, 1993), 59.

[2] See J.J. Colledge, *Ships of the Royal Navy: An Historical Index* (2 vols., Newton Abbot, 1969).

Part One

Introduction

Chapter One

"Guilty Men," Complexes, and Legends

When in 1948 Winston Churchill published the first of his six-volume history of the Second World War, *The Gathering Storm*, he was in no doubt as to how Britain had ended up in such an awful conflict so soon after the Great War of 1914–1918. "*How the English-Speaking peoples through their unwisdom, carelessness, and good nature allowed the wicked to rearm*" was the theme – and the subtitle – of the book.[1] The Second World War, like the First, had cost the British Empire scarcely imaginable lives and driven it to the brink of bankruptcy. In Churchill's mind, this catastrophe had been avoidable. To him, the blame lay squarely at the feet of "guilty men," including Conservative Prime Ministers Stanley Baldwin and Neville Chamberlain, for not stamping out Adolf Hitler's belligerence while Nazi Germany remained weak, and with the National Governments of 1931–1935 and 1935–1939 for failing to adequately prepare British defences for a major war despite repeated warnings of impending crisis.[2]

Churchill had, of course, been marginalised by these administrations, and thus settling a political score was no doubt part of his agenda. Less subjective histories in more recent decades have viewed the actions of these "appeasers" in a more sympathetic light. While rarely apologising for them, the severe political and financial constraints under which policymakers operated, including – but not limited to – the economy, industry, and public opinion have become an enduring part of the interwar narrative.[3]

It is often forgotten, however, that during the 1930s an entirely different and mutually contradictory line of thinking developed. The so-called "Merchants of Death" concept argued that shady arms dealers could influence the direction and timing of armed conflict, that Britain and other major powers had uncomfortably close relationships with armament manufacturers for too long, and that private profit in weapons led directly to war. The concept – some would call it a "myth"

---

1 W. Churchill, *The Second World War, Volume 1: The Gathering Storm* (Boston, 1948), xvii.

2 *Ibid.*, 103.

3 J.P. Levy, *Appeasement and Rearmament: Britain, 1936–1939* (London, 2006), x.

– prompted Royal Commission enquiries, mass petitions with millions of signatures, and even freakish by-election results. Since that point the legend has become bundled with Cold War-era warnings of a "Military-Industrial Complex" and has taken on a life of its own. Today, theories of state-military cooperation retain a powerful allure and have been told and re-told for successive generations of conventional and nuclear weapons.

As far as the interwar period of the "guilty men" is concerned, these narratives sit against a backdrop of profound change for British defence. Exiting the First World War with the largest and most modern fleet on the planet, the economic challenges of the 1920s and the development of tanks, bombers and fighters meant that the face of the private British armaments industry, and its place in the world, was forever altered by 1939. In 1918, most large armaments were made in Victorian-era shipyards and engineering plants. The late 1930s, by contrast, were marked by a proliferation of "shadow" factories, capable of converting from cars or bicycles to the latest generation of aeroplanes in a way unthinkable when Britain began its fateful naval arms race with Germany thirty years earlier. In other words, these "guilty men" and "Merchants of Death" arguments were being made in a world where armaments manufacture, and its relationship to the state, was changing at a breakneck pace.

Nevertheless, despite near-constant attention many crucial questions remain unanswered seventy years after Churchill's work appeared. This book examines two: the role of industrialists and private manufacturers of naval arms in British defence planning, and the successes and failures of the supply planning framework for national security. To an extent then, this is a study of how the military-industrial complex worked in practice, and of where myth met reality. It is also set in large part within the machinery of the British government by examining the planners and organisers in the Committee of Imperial Defence (CID) and the ways that the most important private industrialists interacted with them. Although far from newspaper headlines, the cogs that turned in the CID were directly connected to the top echelons of the Cabinet, to foreign policy, and to the Admiralty. But this is not merely a study of procurement. The book also examines the impact these subcommittees had on naval defence and assesses how successful they were in managing Britain's rearmament drive in the 1930s.

The answers to these questions have far-reaching consequences. The outcome of the bids for contracts could, at least in theory, make the winners rich or see established firms fail, as indeed many did during the interwar period.[4] Similarly, state control of industry might have meant greater efficiency and thus have been of net benefit to the rearmament effort, but equally it might have meant sacrificing free competition in favour of supporting existing relationships, and

[4] L. Johnman and H. Murphy, *British Shipbuilding and the State since 1918: A Political Economy of Decline* (Exeter, 2002), 25.

therefore may have been damaging to the state of British defence. Since these decisions had a direct effect on the composition of the British fleet – and indeed on defences in general – when war was declared, it is the aim of this book to examine how this planning machinery worked, who had the greatest impact on it, and what ultimately were the implications for Britain.

The State of Play

There has been a great deal of attention given to the state's procurement of weaponry and the companies that provide it, returning us once more to the search for evidence of a "Military-Industrial Complex." The term itself grew directly from President Dwight Eisenhower's famous farewell address in 1961 where he warned of the "unwarranted influence" of the shadowy network, arguing that "the total influence – economic, political, even spiritual" was being "felt in every city, every State house, every office of the Federal government." "The potential," he said, "for the disastrous rise of misplaced power exists and will persist."[5] In other words, he was abundantly clear about the danger such a relationship posed to democracy. As a direct result of this warning, the term is now almost always used pejoratively, has assumed a vast and pervasive element of business with tentacles spread throughout policymaking, and has created extra (and unnecessary) military spending. Moreover, it is argued, this spending has led to the enrichment of industrialists – and even politicians – through preferential treatment and *quid pro quo* arrangements in lucrative armament contracts. In short, since Eisenhower coined the term, the characterisation of state-industry relations has been almost wholly negative.[6]

Similarly, the "Merchants of Death" legend, although usually credited to a 1934 book by H.C. Engelbrecht and F.C. Hanighen,[7] has existed in one form or another since long before the First World War. In its most common form the legend portrays shadowy arms dealers stoking the fire between rival factions or nations and then selling guns to both sides in a battle in order to maximise profits, though it has also been extended to include any organisation or individual who profited from selling weapons.[8] Fascination with the idea reached its apogee among the British public in the 1930s during a phase of

---

[5] For an overview of these see J. Ledbetter, *Unwarranted Influence: Dwight D. Eisenhower and the Military Industrial Complex* (New Haven, 2011), 15–18.

[6] *Ibid.*, 6–12. See also T. Hayden and M. Pilisuk, "Is There a Military-Industrial Complex Which Prevents Peace? Consensus and Countervailing Power in Pluralistic Systems," *Journal of Social Issues,* 21/3 (1965), 67–117.

[7] H.C. Engelbrecht and F.C. Hanighen, *Merchants of Death: A Study of the International Armaments Industry* (New York, 1934).

[8] C. Trebilcock, "Legends of the British Armaments Industry, 1890–1914," *Journal of Contemporary History,* 5/ 4 (1970), 4.

heightened international tension when they looked to the League of Nations, rather than to armies and navies, to guarantee security. Accordingly, it spawned numerous publications, protests and petitions, most notably the "Peace Ballot" of 1934–5 which gathered more than ten million signatures demanding a ban on the sale of arms for private profit.[9]

Yet despite often being bundled together, these theories do not sit well with each other. One asserts that the state and industry had an uncomfortably close relationship, while the other argues that arms manufacturers caused wars by playing nations against one another. Moreover, while both notions have been the subject of much research, most investigations have tended to "test" whether such a complex or legend existed by applying existing definitions to historical cases.[10] Thus, research has often been constrained within the narrow bounds of what a complex *should* look like and has failed to examine how relationships between industry and state actually developed.

Naval arms are important here precisely because the British naval arms industry before World War II had several distinctive characteristics that set it apart from the rest of arms production. Warships had been built in Britain for decades before 1918, and shipbuilding was integral to cities like Glasgow, Liverpool, London, Newcastle and Portsmouth between the wars. Unlike artillery, optical munitions or even aircraft which, while technologically complex, were produced in the hundreds or thousands, a major warship was an expensive, often one-off item that consisted of thousands of bespoke components, took several years to build, cost several million pounds, and provided work for entire communities. Even when several ships were built to the same design, they were usually constructed simultaneously in yards many miles apart for reasons of speed; thus, mass production techniques could never be fully utilised. As such, there were relatively few firms that built naval vessels, and they were controlled by a small group of individuals, typically with long histories of supplying ships and components to the Admiralty. The British Empire relied so heavily on such ships that the Royal Navy was never far from the thoughts of officials in Westminster and Whitehall. The private naval arms industry thus stands alone as being equally central to British defence in the years before the Great War and on the eve of the Second World War.

There is a substantial body of scholarship on the Royal Navy, the shipbuilding industry, the private manufacture of arms, and British imperial defence in these years. Nevertheless, work on the *political economy* of state and industry is less prominent. While some scholars have tackled important topics associated with it, including early works like N.H. Gibbs' *Grand Strategy* on rearmament

---

[9] P. Kyba, *Covenants without the Sword: British Public Opinion and Defence Policy, 1931–1935* (Waterloo, ON, 1983), 149.

[10] Hayden and Pilisuk, "Is There a Military-Industrial Complex Which Prevents Peace?" 67–117.

policy, William Hornby's *Factories and Plant*, or M.M. Postan's *British War Production*, they were (and are) mainly studies of high policy, omitting almost entirely the role of businessmen.[11]

These themes and lines of argument were developed by later generations of historians, especially in the 1970s and 1980s when G.A.H. Gordon, G.C. Peden, R.P. Shay, and Clive Trebilcock produced important work on various aspects of the wider organisational questions posed by Gibbs and others. Peden's work on the Treasury helped debunk the myth that rearmament failed because it was "starved" of funds, while Gordon and Shay examined for the first time the procurement process and the relationship between business and industry, bringing industrial figures into slightly sharper focus. Nevertheless, as praiseworthy as these works remain, Gordon's study is primarily a naval history with businesses on the periphery, while Shay's was of rearmament from the top-down and thus afforded little space to the fortunes of businessmen and their interaction with the state's supply planning framework. Trebilcock's research, however, on the arms giant Vickers and on armaments manufacture (confined to the years before 1914) is noteworthy for its direct criticism of what he described as "myths" and "legends" surrounding the Merchants of Death allegations.

Most recently David Edgerton, Paul Kennedy and Joseph Maiolo have reinvigorated a field which had lain near-dormant for two decades with some striking new assessments. These include observations on the nature of the arms race in the 1930s, the "problem solvers" who overcame technological challenges during the war, and a reassessment of Britain's supposed shift from "warfare" to "welfare" state. Edgerton has been particularly active in revising the conclusions of Shay and Correlli Barnett (who argued that British power "collapsed" after the Great War), arguing forcefully against the view that Britain materially declined. Instead, his work gathered much attention – and a good deal of controversy – over his conclusion that Britain remained the world's most powerful and technologically sophisticated warfare state in 1940. Nevertheless, in doing so he injected much-needed life into a topic which had been largely neglected throughout the 1990s and the early part of this century.

This book suggests that for all this revision, the story of state-industry interaction upon which this "warfare state" was built, has not been fully told for the interwar years. Perhaps most importantly, when industry is discussed in the existing literature the focus is too often only on Vickers, the armaments giant based in north-eastern England. In previous studies Trebilcock, Richard Davenport-Hines and others have used Vickers as not only the leading armament firm but also *the* firm from which to draw more general conclusions about

[11] For instance, one of the government's most senior industrial advisers, Lord Weir, is mentioned once, in a footnote, nearly 600 pages into Gibbs' work, but not at all in Postan or Hornby, while another, Sir James Lithgow is not mentioned at all in any of them. Gibbs, *Grand Strategy, Volume 1: Rearmament Policy* (London, 1976), 582n.

private industry.[12] This is perhaps understandable given Vickers' size, history and expertise, but it is misleading in the contexts of the navy and of rearmament. To give but one example, even at Vickers' lowest post-war ebb, it still recorded a yearly profit of over £400,000 and, for all but three years between the wars, paid dividends of four percent or more to shareholders.[13] This was entirely atypical of other armaments manufacturers and, as will be shown, the notion of Vickers' centrality within the industry often does not tally with discussions over armaments policy occurring in the CID.

In sum, the nature of private naval arms manufacture suggests that there is much scope for considering it a special case: intertwined with the other aspects of defence but, due to its size, history and products, nevertheless distinct. These family-owned industries had histories and relationships with the state and Admiralty, stretching back some seventy or more years by 1939. Their directors were powerful and well-connected; their products extremely difficult and expensive to make. Yet histories of rearmament as a single entity (and the disarmament that preceded it) have not always captured the naval peculiarities. This is striking not least because of some of the strong claims made about decline or military-industrial complexes. Their role in defence and the circumstances faced between the wars means that scope also exists for a more detailed study of the industry's relationship with the planning and procurement machinery, and not solely Whitehall's relationship with industry. This is the underlying aim of the chapters which follow. The focus will now move towards describing and defining the supply organisation framework that existed within the British government and to examining how both this and the industry changed in the years following the Great War.

---

12 Trebilcock, "Legends," mentions Vickers forty-one times. The Brown, Scott, Fairfield, Beardmore, Palmer, Armstrong-Whitworth and Coventry Ordnance works are mentioned a total of eleven times. Similarly, E.F. Packard, "Whitehall, Industrial Mobilisation and the Private Manufacture of Armaments: British State-Industry Relations, 1918–1936" (PhD thesis, London School of Economics and Political Science, 2009), 34, mentions the other companies "which struggled next to Vickers." See also R. Davenport-Hines "The British Armaments Industry during Disarmament, 1918–1936" (PhD thesis, Cambridge University, 1979).

13 J.D. Scott, *Vickers: A History* (London, 1962), 390–391.

Chapter Two

Structural Changes, from the Ministry of Munitions to the Principal Supply Officers Committee, 1918–1927

The Structure of Industry between the Wars

In truth, apart from 1919 and 1920, when order books were full to replace vessels lost in the war, the period up to 1935 was mostly a struggle for the private shipbuilding industry. The experience of the naval race with Germany before the Great War had created a set of circumstances that could not last forever and certainly could not be matched in the post-war world. The Scottish shipyards on the River Clyde alone launched one-quarter of the world's tonnage in 1913, a figure which owed much to its strong warship sector.[1] Thus, after the war, the Royal Navy possessed a large, young, and expensive fleet, but faced no obvious enemies following the collapse of the Imperial German Navy. Moreover, the British public had been shocked by the horrors of the Great War and now demanded that elected representatives turn their attention to arms limitation and international treaties to ensure there would be no repeat. As such, continuing the high levels of expenditure on the Royal Navy was not a priority for successive British (or other) governments. Industry was left to face the 1920s with only a fraction of the orders it had previously enjoyed.

In private, the government's thinking was guided by the so-called "Ten Year Rule." Adopted secretly in 1919, the rule assumed that because no major conflict was likely within the next decade there would be no need for a major construction programme.[2] This was renewed annually until 1932, when it was revoked following events in the Far East. But within just five years of its repeal, events in China and later in Germany and Italy set in motion the largest armaments drive in British history, an enterprise in which the Royal Navy played a major part. The problems by 1937 were very different for the British government and

---

[1] C.A. Oakley, *The Second City, Glasgow* (Glasgow, 1967), 170.

[2] D. Richardson, *The Evolution of British Disarmament Policy in the 1920s* (London, 1989), 11.

industry. After a period of uncertainty when the government feared the political and financial consequences of rearmament, it soon found that it could not move quickly enough: shortages of skills and plant were holding up the effort to put the country's defences in a state of readiness, while severe bottlenecks – large guns were a problem, armour plate another – persisted. When viewed together, the swing from war to disarmament and then to rearmament placed great stress on a small government planning organisation and tiny number of firms capable of undertaking rearmament contracts.

In short, the reactions to disarmament and rearmament are two of the main themes of this book. It should not, however, be forgotten that the foundations of the latter were laid out during the phase of disarmament which preceded it. In the 1930s, businessmen, politicians, civil servants and commissioned officers interacted in secret to organise industry and preserve capacity, but they did so using a committee framework developed in the 1920s. For this reason, the successes, failures and profits that came in later years cannot, and should not, be separated from the leaner earlier times.

It is instructive also to outline what the "industry," at least for the purposes of this book, looked like. Despite appearing simple at first glance, historians have nevertheless defined it as everything from "£4 revolver[s] to warships … worth several millions."[3] This was not lost on contemporaries either. In a letter to *The Times* in November 1933, the shipbuilder J.E. Thornycroft argued that all industrial companies in the United Kingdom could be considered as "potential armaments works." Using his own yard as an example, he noted that it was the drive to make "safer and faster boats" that led him to develop the "water tube" boiler; the "efficiency of this development led to its implementation in naval vessels." In his eyes everything from steel to chemicals and even the boots on soldiers' feet could and should be considered weapons of war.[4] The following week, Lord Weir, chairman of the giant engineering firm G&J Weir, echoed Thornycroft by arguing that mercantile and naval ship developments had benefitted each other on numerous occasions and that a list of "armament works" was in effect a catalogue of "the larger proportion of British industrial units."[5] Three years later, the Cabinet Secretary, Sir Maurice Hankey, took somewhat the opposite position in giving evidence before the Royal Commission on the Private Manufacture of Armaments, though he at least agreed that sharp delineations between sectors was unwise. By that point a colleague of Weir, Hankey believed that since no firm was wholly dedicated to the manufacture of war materiel there were therefore "no purely armament firms" operating in Britain.[6]

---

3 Packard, "British State-Industry Relations," 220.

4 Great Britain, National Archives (TNA), CAB60/40, PSO(SB)387, 30 November 1933, "Industrial Mobilisation: Note of letters to *The Times*."

5 *Ibid.*

6 *Ibid.*

All three men had a point. Splitting the "naval" components apart from the private arms "industry" only complicates the task further. Before 1914, the centrality of the Royal Navy made it relatively easy to identify which private firms made their money from selling weapons. Owing to a rapid technological progression and given added impetus by the actualities of the First World War, especially the development and proliferation of tanks, bombers and fighters, the defence picture was very different by 1939. During the Great War, many firms made components for more than one of the defence service departments. Some obvious examples include steel forges, such as Beardmore's in Glasgow, which made armour plate for the Navy's battleships as well as for the army's five-ton light tanks. Ball bearings manufacturers' products were vital both in gun turrets on ships weighing several hundred tons, and later in much smaller aircraft fighter or bomber engines. Less obvious cases include firms that, although engaged ordinarily in work for one service, might turn over capacity to another in response to certain problems, as Weir's firm did in the Second World War for field artillery (it was ordinarily employed in pumps manufacture for, among other things, warship boilers). Shadow factories in the late 1930s are one of the best-known examples of this kind of scheme.

It is because of such factors that any separation of the strictly "naval" components of the industry would never fully succeed. Nevertheless, while overlap inevitably existed and thus there are some unavoidable generalisations, the importance of understanding the history of naval armaments manufacture is paramount to the story of state-industry relations. One way to "pare down" the sector to just the key private naval armaments firms is to concentrate on those which had a high concentration of armament orders. This is not a new approach: Vickers – in terms of capital by far the largest armaments firm[7] – has been used as a proxy for the behaviour of industry before, although taking only the largest and most successful firm is not particularly useful for understanding how the rest of industry reacted to challenges. Moreover, looking for only high concentrations of armament work can be problematic: while naval contracts composed just one-sixth of the tonnage launched by John Brown of Clydebank between the Wars, it made up half of its total profits in the same period. As such, a low concentration of Admiralty orders did not mean a firm was not an important armaments manufacturer.[8]

A more instructive approach would be to use construction "bottlenecks." These existed because certain items were sufficiently difficult and/or expensive to produce that only a select few firms could make them. For example, any shipyard

---

7 A. Slaven, *British Shipbuilding, 1500–2000: A History* (Lancaster, 2013), 58.

8 A. Slaven, "A Shipyard in Depression: John Brown's of Clydebank, 1919–1938," *Business History*, 19/2 (1977) 192, 202; University of Glasgow, Scottish Business Archives (GUAS), UCS 1/74/7–11 and UCS 1/5/19, John Brown and Co, Clydebank, tender books.

in the merchant or passenger sector could construct a hull, but only a few could produce armour plating or high-powered engines, turrets, fire control devices, and other material that set a warship apart from everything else afloat. Indeed, it has been suggested that in 1914 the ability to make a revolving gun turret for a battleship was about as rare as the capacity to make a spacecraft seventy years later.[9] Quite apart from the technology, very few had the skilled labour to fit out such a ship. A typical 8000-ton cruiser, not large by battleship standards, had a manpower cost around twenty times that of an equivalently sized cargo vessel.[10] The *Town*-class of cruisers, for instance, cost a little over £2m from John Brown in 1935,[11] while a similar size cargo ship from the same yard cost £83,500 in 1936.[12]

It was these bottlenecks that dictated much of the financial cost of warships. Some forty percent of the cost of a warship could be in the turrets and guns.[13] Moreover, unlike the building of cargo vessels, where doubling the size had a comparatively small effect on the final cost, a massive battleship such as those in the *King George V*-class was between four and five times larger than a cruiser but – owing to the complexity of components – cost almost four times as much.[14] Even simple escort vessels required five times as many work hours as their mercantile equivalents.[15] Owing to the depressed market conditions after 1920 and the business failures and consolidations that followed in their wake, there were even fewer firms in 1933 capable of attempting such a task compared with 1918 (see Table 2.1 and 2.2, below). Therefore, it is through those problems and issues most critical to the maintenance of Britain's capacity to wage war and protect trade routes that an assessment of political economy is to be made.

Indeed, based upon the location of successful Admiralty tenders between 1918 and 1939, there were no more than fifteen shipyards at any one time in Britain that built vessels larger than 1000 tons (the size of a small destroyer) for the Admiralty or foreign navies. A highly secretive organisation dating from 1926 known as the Warshipbuilders Committee (WSBC) comprised such yards and

9 C. Trebilcock, "Science, Technology and the Armaments Industry in the UK and Europe, with Special Reference to the Period 1880–1914," *Journal of European Economic History*, 22/3 (1993), 569–570.

10 G.A.H. Gordon, *British Sea Power and Procurement between the Wars: A Reappraisal of Rearmament* (Hong Kong, 1988), 188–189.

11 GUAS, Papers of John Brown and Company, UCS1 75/11, cost books. While the total finished cost of the ship was £2.14m, the portion that Brown's constructed comprised £1.07m, or fifty percent. The remainder was for guns, armour, engines, etc. ordered from and built elsewhere.

12 GUAS, UCS1/9/79, "Memorandum on Conditions Now Existing in the Shipbuilding Industry, December 1938."

13 F.E. McMurtie (ed.), *Jane's Fighting Ships of World War II* (London, 1997), 25.

14 D. Edgerton, *Britain's War Machine: Weapons, Resources and Experts in the Second World War* (London, 2011), 212.

15 Gordon, *British Sea Power*, 188.

Table 2.1:
Report on the Condition of the Private Armaments Industry

Company in 1914	Situation in 1933
Coventry Ordnance Works	Out of business
Birmingham Small Arms	Not engaged on armament work
London Small Arms Co.	Out of business
Armstrong Whitworth	Absorbed by Vickers
Cammell Laird	Vastly reduced capacity
Beardmore	Reduced to nucleus. Small orders for naval guns only
Firth	Only small orders for shell. No other arms work
Hadfield	Only small orders for shell. No other arms work
Projectile Co. Ltd	Practically no orders
Darlington Forge Co.	Out of business
John Brown & Co.	No longer do gun forgings
Vickers	Only firm left capable of large orders of all kinds

Source: "PSOC Report: Private Armaments Industry," PSO1109, 31 March 1933, TNA: CAB4/22

thus is an obvious candidate for study. Members were drawn from a self-selecting circle of firms typically engaged in both the highest volume and concentration of domestic and foreign naval orders between the wars, even though most ordinarily were also engaged in merchant or passenger work. For most of its existence the WSBC comprised fifteen companies: Vickers-Armstrong, Beardmore, John Brown, Cammell Laird, Denny, Fairfield, Harland & Wolff, Hawthorn Leslie, Palmer, Scott, Stephen, Swan Hunter, Thornycroft, White and Yarrow.[16] The official reports, Admiralty tenders and membership of the WSBC all point to the same small subset of firms.

The second group of interest here consisted of firms which, although not builders of ships, benefitted from and relied upon Admiralty orders. Here, the definition of "benefit" needs to be narrower than J.E. Thornycroft's view that all firms that profited from arms manufacture were part of the "arms industry," for this broad definition could reasonably include carpenters and outfitters in shipyard towns like Clydebank. Instead, the most useful approach involves looking once again at the high-value items which ships needed but only a small number of firms could make. Certainly, the committees that comprised the supply

[16] D.K. Brown, *A Century of Naval Construction: A History of the Royal Corps of Naval Constructors* (London, 1983), 23–24.

Table 2.2:
Report on the Position of Private Armaments in Imperial Defence

Warships (Hull)	J. Brown & Co.
	Fairfield S. & R. Co. Ltd.
	Scott S. & K. Co. Ltd.
	Yarrow & Co. Ltd.
	Hawthorn Leslie & Co.
	Palmer S. & I. Co. Ltd.
	Swan Hunter
	Vickers-Armstrong Ltd.
	Cammell Laird & Co. Ltd.
	J. I. Thornycroft & Co. Ltd.
	J. S. White & Co.
Warships (Engines)	Wm. Beardmore & Co.
	J. Brown & Co. Ltd.
	Fairfield S. & R. Co. Ltd.
	Scott S. & K. Co. Ltd.
	Yarrow & Co. Ltd.
	Hawthorn Leslie & Co.
	Palmer S. & I. Co. Ltd.
	Parsons Marine S. T. Co. Ltd.
	Wallsend S. & E. Co.
	Cammell Laird & Co. Ltd.
	J. I. Thornycroft & Co. Ltd.
	Vickers-Armstrong Ltd.
	J. S. White & Co.
Warships (Armour)	Wm. Beardmore & Co.
	Firth-Brown (Part of J. Brown)
	English Steel Corporation (Subsidiary of Cammell Laird & Vickers-Armstrong)
Guns	Vickers-Armstrong Ltd.
	Wm. Beardmore & Co.
Gun mountings (Turret)	Vickers-Armstrong Ltd.
Gun mountings (Transferable)	Vickers-Armstrong Ltd.
	Harland & Wolff, Scotstoun

Source: "The Position of Private Armaments in Imperial Defence," PSO359, 10 March 1933, TNA: CAB60/13

organs of the CID were organised in this way. So, for instance, Weir's products were essential not only for marine engines during the Great War but also later in aircraft and artillery, although the group did not produce ships.[17] Though any selection will inevitably exclude firms which ultimately assisted and/or benefitted in some way from rearmament, focusing on the companies and individuals that either dominated their own markets or were involved in producing the goods most difficult to obtain allows for the best assessment of industry-government-military interaction.

To this end, there is one other group worth noting: finance and financiers. Though finance has often been called the "fourth arm of defence," the operations of banking and financial systems during the interwar period have been covered well by Peden, Bowden, Collins, Pugh[18] and others, so this book will instead focus only on instances where the Treasury and Bank of England interacted with planning and industrial experts. For example, the way in which Sir James Lithgow gained financial backing for his exploits in the National Shipbuilders' Security scheme is discussed at length in chapters three and four, while other schemes to assist industry financially have been omitted.

War and Post-war

At this juncture, it is important to understand how the machinery of government worked and developed between its inception before the Great War and the 1920s. The first important milestone came in 1902 when the CID was founded. This was intended to replace the "Defence Committee" – usually only convened in times of crisis – and to be a small and permanent advisory committee to the Prime Minister and Cabinet. Initially devised to define the future strategic priorities of the navy and army, the CID soon took on new duties and powers in the years before the Great War.[19] Under Maurice Hankey, the Cabinet Secretary, the CID gained importance and by 1914 had moved away from just advising the Cabinet – although authority ultimately rested with ministers – to planning and conceptualising defence for Britain and the wider empire.

During the Great War, another highly important body was established. The responsibility for coordinating armaments manufacture was given to the Ministry of Munitions, which was to be officially separate from the CID. Hastily created and given enormous power in 1915 in response to a severe shortage of

17 W.J. Reader, *The Weir Group: A Centenary History* (London, 1971), 75.

18 S. Bowden and M. Collins, "The Bank of England, Industrial Regeneration and Hire Purchase between the Wars," *Economic History Review,* 45/1 (1992), 120–136; P. Pugh, *The Cost of Seapower: The Influence of Money on Naval Affairs from 1815 to the Present Day* (London, 1986).

19 F.A. Johnson, *Defence by Committee: The British Committee of Imperial Defence 1885–1959* (Oxford, 1960), 88.

shells for the front line, the organisation grew to become the largest government department ever seen, with an expenditure by 1918 three times higher than the entire government's budget in 1913.[20] Reflecting this, senior politicians such as David Lloyd George (1915–1916) and Winston Churchill (1917–1919), both future Prime Ministers, were among those who headed the ministry during its short life. Though it was disbanded in 1921, part of its role was taken over in 1927 by new and expanded functions of the CID's peacetime organisation, linking strategy and procurement together under a single body.

The story of the CID's expansion to incorporate the Ministry's peacetime roles is significant, primarily because of the way in which it came about. During the war the Ministry was a nexus – *the* nexus – where government and industry cooperated on industrial planning. Crucially, its staff were drawn from a wide range of sources: civil servants from the War Office and other government departments, judges, accountants and academics were thrown together, often in a great hurry, and ordered to work *under* a top-level drawn from business elites. Indeed, the inaugural minister, Lloyd George, claimed from the outset that the Ministry must be "from first to last a business-man organisation" and that he intended to "utilis[e], as far as possible, the business brains of the community … some of them at my elbow … to advise, to counsel, to guide, to inform, to instruct, and to direct."[21]

Industry responded with enthusiasm to the opportunity to be involved in the war effort. Within a matter of weeks after the Ministry's establishment in 1915, Lloyd George had secured the service of dozens of "distinguished captains of industry" to assist his cause, to which he "entrusted the first position in every department." He gave personal support and authority to these men to break through much of what he called the "aloofness" which characterised the normal administration of government contracts.[22] One of his most senior appointments was William Weir, the thirty-seven-year-old chairman of the Glasgow pump manufacturing firm G&J Weir, which he had inherited from his father in 1910, and in wartime was supplying boilers to the Admiralty. Weir was made Director of Munitions in Scotland, the first appointment in what would be an extremely close thirty-year relationship with the government.[23] By the conclusion of hostilities in 1918 Weir had been first knighted and then elevated to a Baronetcy. Owing to his service and experience within the Ministry, he was chosen in 1922 to head up a committee, known thereafter as the Weir Committee, to report on

20 S. Marriner, "The Ministry of Munitions 1915–1919 and Government Accounting Procedures," *Accounting and Business Research*, 10/1 (1980), 130.

21 D. Lloyd George, *War Memoirs,* Volume I. (London, 1933), 148–149.

22 *Ibid.*, 148–150. Although Lloyd George claims that this speech was made in the House of Commons on 14 June 1915, the author can find no record of this. Most likely it was made the previous week in Liverpool or Manchester to an audience of local employers and workmen.

23 Reader, *The Weir Group,* 75.

the "Amalgamation of Services Common to the Navy, Army and Air Force," whose work might form foundations to base an enquiry into the "Supply of Munitions and Armaments in a Future War."[24]

Given the size, cost and purpose of the Ministry of Munitions, it was without a natural role in peacetime and was soon rendered obsolete. Weir's task, in its simplest terms, was to recommend a way forward towards a system that would be suitable in peace while being able to plan effectively for war without the unwieldy size or bureaucracy of the defunct ministry, and without most of the cost. He did not have complete autonomy; there was significant parliamentary support for a new centralised Ministry of Defence that would take over the supply arrangements of the three services. Opposition was equally vociferous, and spearheaded by the Admiralty, which, as the most powerful of the fighting services, wanted a committee comprised of all three bodies to coordinate supply.[25] While the finer details of this debate are not relevant to the main thrust of this narrative, suffice it to say that in his report Weir broadly took the Admiralty view, arguing strongly against amalgamation and instead for coordination among the common aspects of the three service departments.[26] Two further committees headed by the Marquess of Salisbury and Sir Laming Worthington-Evans followed Weir's, drawing similar conclusions and building upon his recommendations.[27]

Once ratified, this proposed organisation would become in 1924 the Principal Supply Officers Committee (PSOC), later responsible for the important task of supply planning throughout the rearmament phase. The PSOC was initially part of the Board of Trade, but in 1927 was reconstituted as one-half of the CID (see Figure 1), dealing with the supply side, while the other half – the Chiefs of Staff (CoS) – was responsible for assessing and defining threats. The operation of the CoS is of lesser relevance to this story, but broadly speaking the PSOC's work was guided by its views. The three chiefs – Army, Navy, Air Force – essentially decided upon the nature of any external threat to Britain (including location, type of conflict and the time-frame involved), and instructed the PSOC to organise industry and materials based on what would be needed to counter it. In short, the PSOC only gained momentum as an organising body once the CoS believed that an armed conflict was likely and could convince Cabinet of their views.[28] As far as the 1920s are concerned, the important point is that the unification of supply and strategy within the CID happened because a civilian businessman, albeit one connected to

[24] TNA, CAB15/20, Maurice Hankey (Contracts Coordinating Committee), "Supply of Munitions etc.," 28 January 1921. The second part of the question on the supply of munitions was a pre-existing committee headed by Sir W. Graham Greene, but it was suspended pending the outcome of Weir's report.

[25] Johnson, *Defence by Committee,* 167, 176.

[26] Gordon, *British Sea Power,* 35–36. For a fuller analysis of these discussions, see Gordon, chapters 3–5.

[27] *Ibid.*, 35–37.

[28] This did not occur – as is discussed below – until Manchuria in 1931.

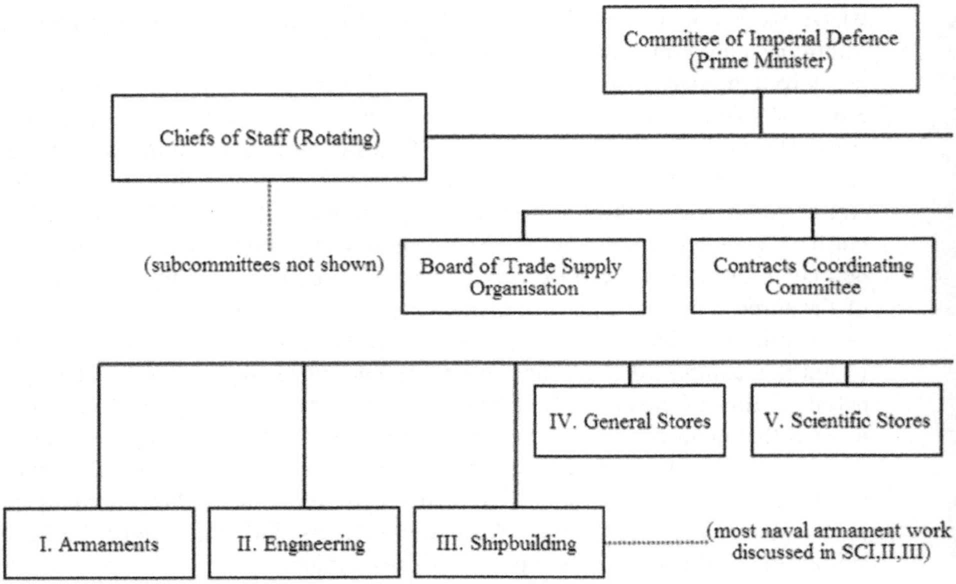

Figure 1: "Organisation for the Coordination of War Supply"
Source: TNA, CAB60/15, PSO(SB)570, February 1936.

the Admiralty and with extensive experience of wartime industrial governance, recommended such a union. Moreover, in addition to being one of the PSOC's forefathers, Weir would, much later, also become one of its chief advisors.[29]

The new PSOC, broadly speaking, aimed to alleviate procurement tension between the service departments by bringing together some of the most senior officers from all services to hold regular forums on future supply needs. It did this through a series of subcommittees responsible for individual items or sectors, such as shipbuilding, which was handled in Supply Committee III. This arrangement suited the Admiralty best and the Army least, and it is likely that the nature of Weir's recommendations may have been to do with the alliances he forged during the war.[30] Regardless, the new committee was to sit atop several sections and subcommittees covering everything from shipbuilding to machine tools and foodstuffs, and was to report to the CID's executive body, which acted as the umbrella organisation over the CoS and PSOC.

As far as governance is concerned, the PSOC – both before and after its reconstitution in 1927 – was chaired by the President of the Board of Trade (another of the positions that Lloyd George and Churchill had both held before

[29] See Chapter Six, below.
[30] Gordon, *British Sea Power,* 36.

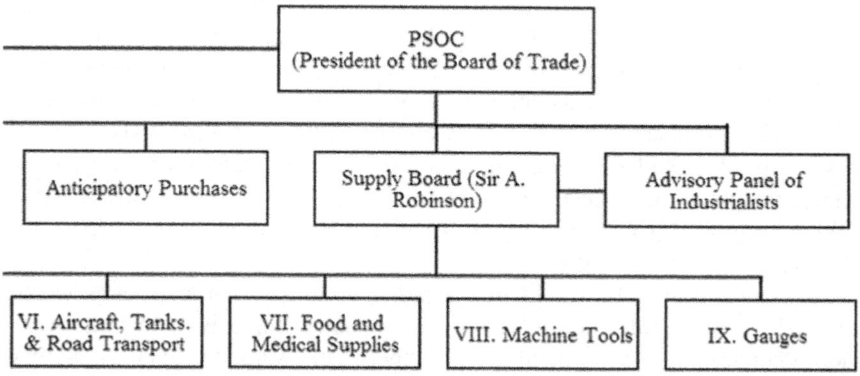

1918). The key figure in the organisation, however, was the government's most senior civil servant, Maurice Hankey. Hankey, a staunch supporter of the navy and sea power, served as both secretary to the CID and the Cabinet – a job so great that it was split into two upon his retirement in 1938 – and he has thus been credited by historians as the "high priest of coordination" and a man of "great significance throughout the two brooding [interwar] decades."[31] His seniority within the government machinery led his biographer to call him a "man of secrets," such was the level of trust in which he was held by successive Prime Ministers and other senior politicians.[32] He oversaw and managed the CID framework in which the PSOC operated, and both consulted and was consulted by the Cabinet on any major piece of work that emanated from any part of the organisation.[33]

31 Johnson, *Defence by Committee,* 166–169.

32 S.W. Roskill, *Hankey, Man of Secrets, 1931–1963* (London, 1974), 22.

33 Within the PSOC, there were two major supply planning organisations: the Supply Board and the Board of Trade Supply Organisation. The former dealt with finding manufacturing capacity for the fighting services (i.e., the navy, air force and army) and was comprised mainly of serving officers and received first call on resources, while the latter dealt with the needs of civilians. The primary difference between the PSOC and its subordinate Supply Board was that the latter had a narrower focus and was not tasked with considering questions of contracts or anticipatory purchases of raw materials, which were handled in separate subcommittees that also reported upwards to the PSOC (see Figure 1). The only exception to the civil-military divide was shipbuilding, where

In sum, the PSOC was formed along the lines of powerful Admiralty interests in the early 1920s, while a large part of its work in later years was related to the navy and shipbuilding. This book therefore tells the story of these cogs in a machine of vital importance, though they were cogs that did not always mesh smoothly. Indeed, critics of the CID and its subcommittees have alleged serious failings and usually characterised it as a scarcely relevant, overly bureaucratic talking shop (it is, for example, cited that even during a typical disarmament year the CID and its subcommittees in the CoS and PSOC met almost six hundred times in all).[34] For the PSOC, most serious among these allegations is that despite talking and planning almost endlessly for years, it did not actually achieve anything tangible during the years of peace, and Britain remained dangerously underprepared on the eve of war.

These claims will be tested in the pages to follow. For now, it is worth emphasising that for defence planning and industry, there was no "interwar" period. Planners did not see themselves as "between" wars. As much as another war was feared in the 1930s, it could never be fully known who the primary enemy would be. Nor was it known whether the battle would range over continental Europe or the British Empire; whether it would take place on land, sea, air or a combination of all three; or whether it would be a limited war or a general conflict that would engulf nations. Hypothesising was therefore of paramount importance. The criticism the CID's bodies have drawn in the historiography, not helped by a PSOC admission in the mid-1930s that the "constantly changing technology and character of armaments production" meant that its work was often quickly outdated[35] is thus unfair. That technology and the geopolitical landscape could shift quickly is hardly surprising. The more pertinent questions – along with how successful the planning framework was – concern how effectively supply planning changed and developed, what role industrialists played, and who benefitted and lost out from it.

It may be thought that the answers are primarily to be found in the 1930s. In some senses this is true, but this does not mean that disarmament policy is irrelevant to rearmament, for the roots of the PSOC and of industrial collaboration with it were sown in the 1920s, as indeed were those which would grow later into the blueprint for rearmament planning. Nevertheless, the most important planning phase occurred in the period immediately after the Japanese invasion of Manchuria and the election of a National Government in Britain,

---

both civil and defence requirements were handled by the Supply Board, under Supply Committee III, chaired by the Admiralty's chief engineer. It was within the individual supply committees and the Supply Board that the industrial planning for defence in peacetime was undertaken and where what one might term a "military-industrial complex" was born. See Gordon, *British Sea Power,* 128.

[34] Johnson, *Defence by Committee,* 221.

[35] For criticism see Packard, "British State-Industry Relations," 94–96. For the statement, see TNA, CAB4/22, PSO 1120, PSOC Report, 6 December 1933.

two events which overlapped with one another in September and October 1931. This stirred Britain – slowly – into action, although powerful constraints on manoeuvre prohibited the increased level planning from becoming policy for several years afterwards. It was these constraints that brought industry into the planning framework, as both had a shared desire to help the armaments industry recover from the effects of the 1920s. While the debates over when and how to rearm that followed Manchuria were coloured by the financial and political pressures stemming from the Great War and the Great Depression, it is hard to conclude that the slow stagnation of armaments until that point would have been shaken in quite the same way had it not been for a new form of government and peril in the Far East. In other words, understanding the stagnation of the naval armaments industry 1920s is crucial to the story of defence planning more generally.

Related Later PSOC Developments

As an organisation interested in supply planning and avoiding bottlenecks the PSOC was, broadly speaking, underworked between its formation in 1924 and the invasion of Manchuria. While conscious of (usually Admiralty-led) warnings of a reduction in skills and capacity through a prolonged lack of orders – especially following the onset of severe recession from 1929 – the committee was not in a position to directly influence policy. Instead, its remit was to tackle the supply question in preparation for a future conflict, which relied on Cabinet and the Chiefs of Staff agreeing that a threat existed. In this regard, nothing changed substantially until Manchuria prompted the revocation of the Ten Year Rule in 1932.[36] While there were certainly geopolitical tensions during the 1920s – most of them concerned, in one way or another, German reparations or arms limitation treaties between the remaining naval powers – from a procurement perspective the strength of British defences was deemed broadly adequate for present needs by the successive Conservative and Labour governments.

Thus, the size and scope of the PSOC's work before Manchuria was limited. For example, it was not until March 1929 that the first tentative report of an "approximate estimate of the total war requirements of the nation in respect of materials" was produced.[37] Even this was rudimentary: the document was only concerned with the largest shipyards, admitted that the capacity of many minor firms was not known, and merely proposed to investigate those questions further at an undetermined later date. The work was suitably titled "A Rough Estimate." There were, nevertheless, several useful reports before 1932 which helped form

---

36 Gibbs, *Grand Strategy,* 75.

37 TNA, CAB60/35, PSO(SB)53, "Rough Estimate of Quantities of Materials Required by Supply Committee III," 5 March 1929.

the basis of future investigations into industrial capacity, prompted by the severe economic depression and London Naval Treaty of 1930.[38] It is fair to say, however, that the content, detail and pace of planning increased following the revocation of the Ten Year Rule: from an initial position of knowing little apart from the capacities of the largest firms, by 1935 the organisational structures, manufacturing processes, supply chains and almost everything else one might conceivably need to know about a defence firm had been documented.[39] This knowledge was extended not just to smaller shipyards but also to automobile and light engineering firms, the chemical industry, and electronics. There was one more striking difference: the PSOC now had a team of civilian businessmen assisting with its top-secret planning.

The invitation that was extended to the industrialists to aid supply organisation is an overlooked, largely unknown, but critical story in the development of defence planning. Their integration into the fold came only in 1933, although the roots of the decision can be traced to the naval threat the Japanese posed by their actions in South East Asia some fifteen months previously. This group owed its existence in some way to the unique domestic political situation created in the wake of the October 1931 general election, for what followed was a period where planning was finally taken seriously but was kept secret, for outwardly the National Government was still committed to peace. The creation of this "Advisory Panel of Industrialists" which – at least initially – reported to the PSOC, and later to the Prime Minister, played a major role in supply planning and represented a peace-time first for state-industry collaboration of this nature. Astonishingly, its existence goes unmentioned in the standard history of the CID.[40]

Instead, the official history focuses on another committee, the Defence Requirements Committee (DRC). The DRC was a committee of Treasury, Foreign Office and fighting services members that advised the Cabinet on defence, but it was concerned with Britain's defensive priorities and not with the supply bottlenecks that pre-occupied the lower orders of the PSOC. It also had no business representation. Formed within a month of the Advisory Panel, the DRC has been called the "beginning of rearmament," although conceptualising it in this way omits consideration of the planning – discussed in chapter five – that took place in the two years between Manchuria and its formation, as well as the role industry played in it. Moreover, in terms of industrial bottlenecks, only focusing on top-level policy from the starting point of late 1933 neglects the changes, acquisitions and problems in industry during disarmament and rearmament,

---

38 See chapter five, below.

39 See, for example, TNA, CAB60/44, PSO(SB)591, "Memorandum by Chairman, Supply Committee III," appendix 1–3, 28 April 1935.

40 Johnson, *Defence by Committee.* Lord Weir is also completely omitted from the narrative.

and thus ignores the story of how industry and industrialists responded to the challenges of disarmament in the 1920s, how the PSOC organised naval supply in the 1930s, and with what success.

The deliberations and workings of these committees and the connections made (and used) by their key figures in the 1930s are the focus of the later chapters. For now, attention must be turned back to industry and the Admiralty in the 1920s, for it provides the context for the later concerns over the inability of naval arms manufactures to meet demand and introduces the industrial element that has been so often overlooked. Simply put, the decisions made in the 1920s and their impact on Britain's capacity for naval construction were a major reason why businessmen were required to help tackle the question of efficiency after Manchuria. Thus, the events in this decade cannot be overlooked.

Part Two

Industry and the Navy before Manchuria, and the Establishment of the Warshipbuilders' Committee, 1919–1931

Chapter Three

From Boom to Bust: The Private Naval Arms Industry and the Admiralty, 1919–1926

The 1920s began for industry with a wave of optimism about the future for global trade and a construction boom as markets reopened and ships of all kinds lost to the war were replaced. The Admiralty also had reason to be optimistic: Britain still possessed the world's largest navy, it was preparing to celebrate the launch of its new flagship and largest ever vessel, HMS *Hood,* and had awarded a small – but significant – number of new construction contracts to private yards, including four for new large battlecruisers. This naval work, coupled with the boom in merchant and passenger construction, amply filled the gap left by the conclusion of emergency wartime programmes. Indeed, even in the absence of wartime orders, the naval arms manufacturers managed near-record levels of profit, output and employment into the second half of 1921.[1] This prosperity, however, was short lived. Thereafter, the outlook changed dramatically. For industry, the increase in global trade failed to live up to the wildly optimistic forecasts, leading to significant supply overcapacity in the merchant marine and the subsequent drying up of further ship orders. For the Admiralty, the Treasury's will to cut costs was forcefully imposed upon it: the desire for a tightening of the government purse strings led to international agreements towards arms limitation which abruptly cancelled even the modest naval replacement work in hand.

Part Two of this book examines the relative fortunes of the British naval arms industry and the Admiralty from 1920 until the formation of the National Government and the Japanese invasion of Manchuria in the second half of 1931. The overriding theme is of decline: shrinking financial resources brought about by war debts and the disastrous return to the Gold Standard led to strict controls on expenditures and exacerbated the declining health of the private naval arms industry. The core argument is that the main developments in the relationship between, and relative status of, the Admiralty and industry in this period can be viewed to a large degree as a series of responses to crises, either the Treasury's financial controls or the collapse of global trade. These responses

1 Johnman and Murphy, *British Shipbuilding and the State*, 11.

were not uniform or uniformly successful. Nor were they always in the "public" sphere – a large part of what followed Manchuria did so because of the layers of secrecy built up in the second half of the 1920s by the state and industry acting independently of one another.

The narrative of these responses is also necessarily non-linear: there is some chronological and thematic overlap with the post-Manchurian period, especially in the case of the National Shipbuilders' Security (NSS) scheme, which ran from 1930. For this reason, the role of the NSS and other developments after 1931 are dealt with in the next section, while the background to it is covered here. Similarly, the development of the CID and the PSOC – including Lord Weir's role – has already been discussed in Part One, while the story after Manchuria follows in Part Three. As far as the 1920s are concerned, however, the PSOC is not a major part of the history of private manufacture – the structure of supply planning developed in these years, but the two did not overlap until Manchuria made a future naval conflict seem far more likely. Instead, it was the Admiralty that spent the 1920s arguing for more attention to be paid to the health of private manufacture.

Most fundamentally, it will be shown below that it was the industry and the Admiralty's shared experiences of "failure" – failure to adapt, failure to convince the Cabinet and Treasury of their views – in the first half of the decade that explain the plans that were hatched in the second. Indeed, one scheme, which came to be known as the Warshipbuilders' Committee (WSBC), was a stunning example of the private naval arms industry's ingenuity. What started as a defensive reaction to a crisis grew into one of the most successful price-fixing cartels in the history of armament manufacture and was one of the most astonishing groups to emerge from the disarmament period before 1931.

Structural Problems in Industry, 1920–1923

After the conclusion of hostilities in 1918, the private naval arms "sector" was comprised of fifteen major firms which among them constructed the overwhelming majority of wartime contracts for the Admiralty, from hulls to guns, engines and armour. Each firm could, and did, make other types of vessels or other products for use outside of the naval sphere in peacetime (in the same way that smaller firms assisted the specialist armament manufacturers in wartime), but for the reasons outlined below this was not always a profitable or viable avenue.

In 1918, the large and medium-sized firms – which all predated the war – were Vickers (Barrow and Tyneside), Armstrong-Whitworth, Palmer, Hawthorn Leslie, Swan-Hunter (based in the North East of England); John Brown, Fairfield, Beardmore, Scott, Stephen, Yarrow (on the Clyde), Cammell Laird (of Birkenhead), J. Samuel White (East Cowes), Thornycroft (Hampshire) and the Coventry Ordnance Works (Coventry). The latter made artillery and guns, while the other fourteen constructed ships of various sizes and specifications. Not every yard could make every kind of warship or indeed even an entire ship. Thornycroft, White and Stephen specialised in smaller vessels (destroyers, corvettes and frigates), while only Beardmore, Armstrong-Whitworth and Vickers among the shipbuilders had in-house gun-making facilities.[2] For ease of description and to avoid unnecessary complexities, when the "private industry" is discussed in the pages to follow it is these firms which are being referred to.[3]

In the first two-and-a-half years of peace after 1918, the broader shipbuilding industry experienced "boom" conditions which had been sparked by the rush to replace ships lost during the war, and had been further enhanced with the reopening of trade routes curtailed since 1914. Shipbuilders and shipowners alike anticipated placing orders for and constructing new vessels to meet the demand for some time to come. The resulting strong demand for merchant vessels amply compensated for the gap left by the cancellation of wartime naval orders.[4]

Merchant orders and naval orders were not the same thing, however, and the Admiralty refused to be overly optimistic. By 1922, numerous prominent Admiralty figures had expressed grave concerns about the future health of the warship builders should the current conditions be left unchecked. Two particularly notable voices were David Beatty, the First Lord of the Admiralty, and Eustace Tennyson D'Eyncourt, the Director of Naval Construction (DNC). In 1920, Beatty warned that "specially skilled labour, accustomed to special warship work, is being dispersed [and] the longer warship construction is put off, the more difficult it will be to find suitable labour," while D'Eyncourt, the DNC from 1912 to 1924 and the man responsible for overseeing warship design, made the same point, arguing that "the total cessation of [naval] construction would involve us in a serious deficiency of trained shipbuilding staff and mechanics."[5]

Outwardly, though, there appeared to be good reason for the government to take such prophesies of doom with a pinch of salt. Some significant Admiralty construction and modernisation work remained in hand until the end of 1921: HMS *Hood*, the navy's largest vessel and new flagship, had just completed the fitting out process at John Brown's Clydebank yard,[6] and some modest replacement and repair work was being undertaken by private yards in Birkenhead, Clydeside and the North East of England. Furthermore, Brown, Beardmore, Swan Hunter

---

[2] I. Buxton and I. Johnston, *The Battleship Builders: Constructing and Arming British Capital Ships* (Barnsley, 2013), 174–176.

[3] The one remaining caveat is that some firms collapsed, merged or changed form in the years between the wars, so the list of constituent firms is not always constant. This said, Harland & Wolff of Belfast was the only addition to an otherwise shrinking list before 1939.

[4] Slaven, *British Shipbuilding,* 71.

[5] Gordon, *British Sea Power,* 76–77.

[6] Buxton and Johnston, *Battleship Builders*, 25–27.

and Fairfield (all future members of what would become the WSBC) received contracts for battlecruisers in 1920.[7] The fleet was, however, both young and large – out of over three million tons of vessels in 1920, half had been built since 1914.[8] Moreover, the geopolitical outlook – with the removal of the German navy as a threat – appeared stable, thus lessening the need to sustain a fleet (and by extension, an industry) at 1918 levels.

Thus, by early 1922 clear signs were emerging that a long-term production boom would not occur. When the growth in international trade necessary to sustain new merchant construction failed to materialise, enthusiasm evaporated. The effect was a nine-fold increase in unemployment in less than three years, with almost a third of shipyard workers out of work by the end of 1923, compared with near full-employment in 1920.[9] The roots of these problems are not difficult to pick out. Financially, Britain was in a precarious position. Forty percent of total government spending was swallowed up by interest payments on the war debt alone (this would continue for the rest of the decade), while the level of reparations – initially intended to help the Allies meet the cost of the war – that could be extracted from a defeated Germany fell far below even the most pessimistic estimates.[10]

This all had a considerable effect on the defence budget. Although the navy's share of spending remained relatively constant before 1931 when compared with the Army and Royal Air Force (see Table 3.1, below), it was cut almost in half between 1920 and 1921, with another one-third reduction the following year, leading to the cancellation of several classes of ship. Moreover, wages and thus expendable income were either not rising rapidly or falling on both sides of the Atlantic. This hit purchasing power in two of the world's largest economies (and trading partners), slowed consumption, and stagnated trading. All of this had a knock-on effect on demand for passenger liners and cargo vessels. Thus, when the British government tightened its belt, the private sector was unable to fill the gap left behind.

Exacerbating the problem of fewer orders after 1923 was the relative increase in shipyard capacity. Compared with 1913, British industry had added almost 200 extra slipways, while almost thirty new firms had been established since the beginning of hostilities. At its wartime peak, naval orders accounted for an estimated forty percent of the workload, and without new naval demand dozens of slipways had become idle in the first two years of peace. In addition, the specialist naval manufacturers were, in the absence of Admiralty contracts,

---

[7] Johnman and Murphy, *British Shipbuilding and the State,* 18.

[8] TNA, ADM 181/101, Navy Estimates, 1919–1920.

[9] Johnman and Murphy, *British Shipbuilding and the State*, table 3; and Slaven, *British Shipbuilding,* 85.

[10] The reliance of the British economy on German prosperity and the effect of the Ruhr Crisis is best expressed in D.G. Williamson, "The Ruhr Crisis, 1923–1924," *British Journal of International Studies*, 3/1 (1977), 70–91.

forced to compete for orders with firms exclusively engaged in the construction of merchant and passenger ships, a situation which compounded the problem. Moreover, while the export of British-built ships to foreign owners had accounted for around a quarter of pre-war sales, by 1923 the industries in Sweden, Japan, Denmark, Norway, Holland, France and the United States had all developed and expanded significantly compared with 1913, cutting off an important income stream for Britain.[11]

Table 3.1: Naval expenditure, 1920–1931

Year	Naval Expenditure (£m)	Share of total defence expenditure
1920	156.5	22.6%*
1921	88.4	30.3%*
1922	80.8	42.6%
1923	56.2	45.5%
1924	52.6	42.4%
1925	53.6	44.8%
1926	59.7	46.9%
1927	57.6	46.4%
1928	58.1	46.4%
1929	56.9	47.7%
1930	55.8	47.0%
1931	52.6	45.0%

Source: Derived from B.R. Mitchell *British Historical Statistics* (Cambridge University Press, 1988)
The figures for 1920 and 1921 are skewed because of the costs of army demobilisation, which accounted for £395m and £292m respectively.

The effect of this was dramatic. Not only was worldwide production lower in 1923, but Britain's share was just two-thirds of its pre-war level.[12] Put another way, the British shipbuilding industry faced fiercer competition at home (from new or expanded yards and firms re-entering the merchant and passenger market after years of mainly naval work) and abroad, all in a climate of weak global trade. The result was a blend of a massive global fleet of almost brand new

11 Slaven, *British Shipbuilding*, 67 and 69–71.
12 *Ibid.*

merchant vessels but comparatively little with which to fill them; thus, there were very limited prospects of orders for the foreseeable future.

These events seemed to support the fears that Beatty and D'Eyncourt had articulated. It is, however, important to emphasise that the issues at the heart of their arguments went beyond this. Beatty and D'Eyncourt were, in fact, highlighting the profound differences between specialist naval manufacture and yards ordinarily engaged in only passenger and merchant ship construction. Beatty's use of the term "specially skilled" is important: he was not alone in the 1920s in being highly sceptical, if naval yards collapsed, of the ability of the remaining non-specialist yards to adapt in future years to the demands of a sudden flurry of naval orders.

This position would guide Admiralty thinking on industrial policy for the rest of the decade. Since the 1880s Britain had been committed to the "Two Power Standard," whereby it would sustain a navy as large as her two nearest rivals. This policy was expensive, but the drive to stay ahead of rivals brought, among other things, innovations such as HMS *Dreadnought* in 1905 – then the most powerful ship afloat. Thereafter, to win a new and intense naval arms race with Germany, British naval construction had continued apace towards and then throughout the Great War.[13]

As a result, private yards had been able to secure contracts with a frequency that allowed the retention of labour and the renewal of machinery for more than a third of a century before 1923. This did not mean that there were no industrial recessions during this time – market cycles continued – but these were short and comparatively mild, in the main because the armaments manufacturers could afford to be less reliant on merchant work.[14] There had been no prolonged period where the specialist naval yards had been faced with the possibility of survival based almost solely on a diet of merchant and passenger work – assuming of course that any work of this kind existed. Specialist skills that had been sharpened in the latest technological developments in the context of a more-or-less regular flow of work ran a real risk in the post-war world of being lost altogether.

A major part of Beatty and D'Eyncourt's problem therefore involved technology. To be fast, manoeuvrable and well defended from the latest generation of sea and air threats, battleships required the most powerful engines, heavy armour, and guns capable of firing the largest calibre shells distances of more than fifteen miles.[15] Building these vessels and sustaining these capabilities required constant investment in types of plant unlike that seen in other yards. They also required skilled manpower and research expenditure to maintain steel forges,

13 Buxton and Johnston, *Battleship Builders,* 8.

14 H. Peebles, "Warshipbuilding on the Clyde, 1889–1939, A Financial Study" (PhD thesis, University of Stirling, 1986), 72–73. Peebles' thesis forms the basis of a book with a similar title. Nevertheless, because the thesis contains some extra detail in places it has been cited separately.

15 Buxton and Johnston, *Battleship Builders,* 154, 168 and 211.

engine factories and huge gun-mounting pits. Merchant and passenger ships had, rather obviously, no need for these facilities or skills, which meant that large parts of naval firms could become obsolete even during otherwise prosperous years of merchant and passenger work. Without the capability to make these specific components, a warship was essentially useless, and naval construction would be heavily constrained even if the general "health" of the industry was otherwise good. Neither Beatty nor D'Eyncourt (or anyone else) had any firm idea of what the full extent of the damage to skills and facilities might be, of course, but both men had good reason to suspect that the lack of naval orders would not pass quickly.

The second part of D'Eyncourt and Beatty's industrial problem was that unlike the casual observer, both of these senior naval figures had been conditioned by the preparations for – and subsequent outcome of – the Washington Naval Conference of 1921–1922, which set out the ground rules for naval construction throughout the 1920s.[16] Moreover, while the state of the market and the competition at home and abroad were grave concerns to the shipbuilding industry as a whole, Beatty and D'Eyncourt's concerns were valid *regardless* of the conditions in the wider market, as they applied explicitly to a much smaller subset of firms which had been previously engaged in supplying a very special kind of ship to a single customer – the Admiralty.

The Five Power Treaty and British Naval Construction, 1920–1923

To understand the Admiralty's concerns and to understand the unique picture for warship manufacturers in the 1920s, it is necessary to examine briefly the Washington agreement – or Five Power Treaty – which set out international agreements between the major naval powers for future construction. It was conceived during a period in the immediate aftermath of the Paris peace conference in 1919 where, despite a lack of credible threats in the wake of the Great War, a naval arms race like that between Britain and Germany before 1914 seemed likely to recur. The two potential belligerents this time were the United States and Japan, behind Britain the second and third largest sea powers. The U.S. Navy was in the process of an extensive fleet modernisation plan, started in 1916, while the Japanese had begun to respond to American expansion with its own programme of eight large ships.[17] Both dwarfed Britain's modest construction plans and, taken together, threatened British ability to defend itself at a future date in the Pacific and Far East, and even threatened the supremacy of the Royal Navy itself.

---

16 Richardson, *Evolution of Disarmament Policy*, 13–14 and 26.

17 C.M. Bell, *The Royal Navy: Seapower and Strategy between the Wars* (London, 2000), 8.

On the other hand, the Treasury argued that Britain could ill-afford to pay for an extensive fleet modernisation like that of the United States or Japan. In 1914, British debt had stood at around one-quarter of gross domestic product (GDP). The cost of waging the war meant that by 1919 the ratio of debt to GDP was nearly six times higher than the pre-war level – and three times higher than the comparable 1919 figure in the United States – and rising.[18] Along with this, the demand for steel during the war had pushed its cost per ton up to three-and-a-half times the 1914 level, which vastly increased the cost of all kinds of ships.[19] With Britain struggling to maintain its lead over its rivals, both the government and the Admiralty had a strong desire to curb the level of construction abroad and embraced talks for future arms limitation.

A conference on the naval question was called for late 1921, and an agreement was reached three months later after many rounds of intense negotiations between that November and the following February. The conduct of the conference and the deliberations themselves are well-known and of limited importance to this narrative, although the ultimate conclusions reached should be noted, for they influenced naval construction until the second half of the 1930s. In short, the agreement crucially allowed parity in capital ships for the first time between the Royal Navy and the U.S. Navy. Furthermore, the size of the Japanese navy was formally set at sixty percent of the British or American level (in other words in a 5:5:3 ratio), with the Italian and French navies much smaller again. Smaller ships were capped at 10,000 tons each but were not subject to the same restrictions.[20] In theory, this preserved Britain's place as an unsurpassed naval power and curbed unrestricted construction of large vessels, two key British aims. While the Admiralty wanted limitations in principle, however, Britain ended up sacrificing a great deal to get it: a ten-year "pause" in the construction of new capital ships (excepting HMS *Nelson* and HMS *Rodney,* which were to be allowed under the terms of the Treaty). Moreover, in practice this new ratio meant a significant reduction in the gap between Britain and her principal rivals, the United States and Japan. The Treaty remained in force until the end of 1936.[21]

Until Washington there had been a scarcely challenged belief in politics and industry alike that despite financial pressures Britain would not concede naval supremacy to the United States and indeed might even consider further construction to at least maintain parity on the seas through the building of even larger battleships.[22] But while the restrictions agreed at Washington secured the

---

18 B.R. Mitchell, *British Historical Statistics* (Cambridge, 1988), 601–603.

19 Slaven, *British Shipbuilding,* 70.

20 Gordon, *British Sea Power,* 70.

21 E.H.W. Tennyson D'Eyncourt, *A Shipbuilder's Yarn: The Record of a Naval Constructor* (London, 1948), 140–141; and G.C. Peden, *British Rearmament and the Treasury,1932–1939* (Edinburgh, 1979), 9.

22 These were known as the "G3" class of battlecruiser. See N.J.M. Campbell

minimum of parity that Britain desired, it came with industrial costs. Britain's pre-conference position as the world's largest navy meant that she had, to all intents, already reached the agreed limit. Thus, while the United States and Japan would be allowed to continue to build up their navies to the agreed ceiling, Britain had to call a halt to construction and wait for her rivals to reach the Treaty's levels. Thus, a near-total stop to warship orders was called.[23] Although this served the Admiralty's goal of capping American construction before the U.S. Navy overtook Britain as the world's largest force, which Beatty and D'Eyncourt accepted would happen, concerns remained over the lack of replacement and refit work – essentially scrapping or refurbishing existing ships before the end of their useful lives – to keep industry in a state of readiness. With the "pause" in construction of new battleships, and relatively few older ships needing imminent refits, this left only smaller vessels, such as cruisers and destroyers, to fill the gap for the next decade.

In financial terms, the Treaty was certainly good news for the Treasury. The large battlecruiser contracts were an immediate casualty, followed soon after by cancellations or heavy modifications to almost every other class of ship on order. The value of the remaining outstanding orders in hand in British yards six months after the signature of the Treaty stood at £721,000. Before negotiations began, it was nearly £12m. In other words, ninety-four percent of orders by value had been struck from the balance sheet.[24] Even this figure underrepresents the scale of cancellations, for many more contracts had been discussed but not yet commenced, and as such were not classified as work "in hand."[25] The result was that rather than even the trickle of contracts that the previous modernisation programme would have provided and which were necessary to keep even a few of the private yards actively engaged in construction dried up, a situation exacerbated by the fact that any capital ship order placed in 1922 would have provided work for one yard until 1926. For the warshipbuilding sector, the Washington agreement was like turning off the tap.

Why then had leading industrialists in the private naval arms sector not foreseen such a scenario in the way Beatty, D'Eyncourt and others had? In this case, it should be remembered that naval arms manufacturers had plenty of previous experience with (and indeed had come to expect) cyclical market fluctuations. The nature of naval procurement was that a glut of orders might materialise in a short space of time, as in the first half of the 1890s or during the naval race with Germany after 1908, and then be followed by a leaner period. Similarly, the cancellation or modification of orders part-way through a contract

"Washington's Cherrytrees: The Evolution of the British 1921–1922 Capital Ships," *Warship*, 1/3 (1985), 44.

23 Johnman and Murphy, *British Shipbuilding and the State*, 18.

24 *Ibid.*, 19.

25 Campbell, "Washington's Cherrytrees," 43.

in response to changing political, strategic or financial situations was not unheard of, and had been a feature of naval construction for decades before 1918.[26]

For these reasons, there was a broad acceptance in the industry that the unprecedented Admiralty construction programme of 1908–1918 would not be repeated and that leaner times would follow.[27] The industry, however, was still guided by the belief that Britain's history as the pre-eminent naval power would ensure that it remain ahead of its rivals, especially the United States, and that there would be a *general* continuance of post-war construction to this end, even if only through replacements rather than fleet expansion.[28] Moreover, few industrialists on the eve of Washington expected a decade-long strangling of naval orders and a slump in both the merchant and passenger markets.[29]

It was these beliefs which had to be quickly revised in the immediate aftermath of Washington. Yet since very few courses of action to revive business activity existed, at least for the naval firms with more specialist product portfolios, it was thus not wholly irrational for this sector to pin its belief on securing some naval orders in the medium-term.[30] The lesson here is that Washington showed that information sharing between private manufacturers and either the Admiralty or Government did not function effectively in the first half of the 1920s. The impending resolution to wipe out orders for many years did not reach the boardrooms of the major shipbuilding companies almost until the ink was dry on the Washington agreement, and the decision was not fully accepted for some years after that. Indeed, some were still anticipating at least the completion of the orders that had been placed for the *Hood*-sized "G3" battlecruisers in 1921 and still hoped for replacement programmes afterwards.[31]

This is surprising given that the Admiralty had been well aware of the possibility of order cancellations for some time before the conclusion of the treaty.[32] For instance, the debate with the Treasury over future expenditure had been raging in private since 1920 – before further orders were placed – and it was only brought to a conclusion in 1922 after a committee chaired by Winston Churchill had been formed to discuss the subject.[33] It is especially odd given the relatively close personnel links between the Admiralty and private industry. Before taking up his role as DNC, D'Eyncourt had worked for many years at both Fairfield and Armstrong's, and a transfer of staff in both directions between the

26 Peebles, *Warshipbuilding on the Clyde,* 68–69; K.O. Warren, *Steel, Ships and Men: Cammell Laird, 1824–1993* (Liverpool, 1998), 163.

27 D'Eyncourt, *Shipbuilder's Yarn,* 129.

28 Johnman and Murphy, *British Shipbuilding and the State,* 19.

29 A. Grant, *Steel and Ships: A History of John Brown's* (London, 1950), 71–75.

30 D'Eyncourt, *Shipbuilder's Yarn,* 129–130.

31 Campbell, "Washington's Cherrytrees;" Peebles, "Warshipbuilding on the Clyde," 182.

32 D'Eyncourt, *Shipbuilder's Yarn,* 129–130.

33 Bell, *Royal Navy,* 21; and C.M. Bell, *Churchill and Seapower* (Oxford, 2013), 97.

Admiralty and private business was not unusual before or during the Great War.[34] Indeed, pacifist critics later alleged that this reflected an unhealthy relationship that was *too* close and that put Britain on the path to war.[35] In this case at least there appears to be little evidence of such closeness.

Why this could occur is therefore not fully clear. The answer lies in some small part in the post-war political and economic landscape and the post-Ministry of Munitions structure. The new and developing supply framework within the CID – with the PSOC not coming under its remit until 1927 – cannot be readily mapped onto the ordering process during the Great War, when, as D'Eyncourt put it, "anything could be called for, anything ordered, and all [had] to be ready in the shortest time possible"[36] or indeed the naval race before it. The last time there was anything like a "normal" pattern of peacetime ship replacement was the pre-*Dreadnought* era, which bore little resemblance to Britain's strategic and financial position in the early 1920s. By far the larger part of the problem, however, was that the Admiralty, like industry, was both naïve and in denial in the years before and after Washington.

To understand fully why shipbuilders were left to react without warning to such major changes, one needs to understand that the Admiralty, like industry, had failed to plan seriously for a world without a regular ship replacement programme. It was not until three years after Washington – when it was abundantly clear that merchant and passenger work had not rebounded and when the warshipbuilding sector had failed to diversify into other lines of business – that the Admiralty sought to assist industry directly. Thus, as the section below shows, the experience of Washington had a significant impact on the relationship between the two formerly close partners.

Another question around the Treaty concerns why the government was so quick to ignore the warnings of senior figures at the Admiralty and not continue with even a modest replacement scheme to assist the industry. Here, one should focus on the immediate aftermath of Washington and particularly the Admiralty-Treasury debate over naval expenditures which would be a recurring theme for the next fifteen years. In 1921, the Liberal-Conservative coalition government was desirous of concluding an arms-limitation deal with the United States and Japan as quickly as possible to curb spending on an arms race that might have spiralled out of control and potentially led to another global conflict. This was compounded by the German situation with reparations and the trouble around reducing war debts to the United States. Both factors heavily constrained Britain's ability to pay for fleet renewal.

---

34 D'Eyncourt, *Shipbuilder's Yarn,* 59.

35 D.G. Anderson, "British Rearmament and the Merchants of Death: The 1935–1936 Royal Commission on the Manufacture of and Trade in Armaments," *Journal of Contemporary History,* 29/1 (1994), 19–20.

36 D'Eyncourt, *Shipbuilder's Yarn,* 129.

If this medium-term outlook contained no imminent military or naval threats, the British government's decision to concentrate on cutting its own cloth seemed a financially prudent move. By the same token, if future expenditure was not to be made available for defence, both the Admiralty and the government believed it necessary to extract guarantees from the United States that the opportunity to surpass British supremacy on the seas would not be exploited. Thus, the case for settling the question with the United States was obvious. Moreover, the Admiralty had little bargaining power for spending within a general foreign policy that was moving in a very different direction than before the Great War.

The fear of being unable to keep pace with the U.S. Navy's spending and thus wishing to agree upon "parity" is not, however, the entire story when it comes to the Admiralty's relationship with the Treasury, with whom it negotiated the finer costs and details of its proposed schemes. This relationship, which grew more fractious as the 1920s progressed, had a knock-on effect on industry and helps to explain much about the way firms behaved from 1926 onwards.

This change came about relatively swiftly. The Admiralty had essentially gotten its wishes during the war and in the re-tooling boom that followed. In 1920, for example, when it was still hoped that world trade would rebound strongly, the British government had made a public commitment in Parliament to ensuring that the Royal Navy "would not be inferior in strength to the Navy of any other Power."[37] This basic fact did not change with Washington. In theory, this was a downwards revision of the "two power" standard employed before the Great War, where the Royal Navy was to be sustained at a level capable of engaging the next two largest navies simultaneously, to a "one power" standard. Though formally consenting to keep pace with a fleet (in this case, the United States) that was, along with Britain's own, far and away ahead of every other on earth, and the one they were arguably least likely to ever face in battle, gave the Admiralty considerable authority – and by extension grounds for seeking increased spending allocations – to construct warships in response to any movement from across the Atlantic. In a world where the fleet was by any measures large, modern, and defended an island nation relatively free from external threats, testing the commitment of Parliament to match Britain's Atlantic neighbour could have been a very expensive undertaking.

From this starting point, the balance of power in the Admiralty-Treasury relationship shifted as the latter sought to reaffirm its control over spending. Parity with America was no longer enough; the Treasury argued for definite limits on construction and fleet size. The motivation was as much about wanting to wrest control of policy and spending from the Admiralty as it was about any rivalry with the United States. As a result, a clause was inserted in the Washington Treaty that agreed not only on parity but also to a fixed tonnage (of

[37] Hansard, House of Commons Debate [hereafter HC Deb], vol. 126, c.1210, 17 March 1920.

a little more than half a million tons of capital ships) for the fleet and a defined total number of ships to be maintained.[38] As C.M. Bell has argued, this "negated any earlier victories the Admiralty had won" on matters of construction over the Treasury between the end of the war and the conclusion of the conference in February.[39] At a stroke, it removed much of the authority the Admiralty had on matters of naval construction. The outcome of the Treaty therefore marked a watershed and served as one-half of a vicious double blow to the private arms manufacturers. Not only were prospects for naval construction badly hampered but this was occurring against a backdrop of weak global growth and extremely weak demand for merchant and passenger vessels.

Except for two new ships, there was to be a ten year "pause" on new construction.[40] On one hand, given the relative youth of the fleet in the 1920s (the *Queen Elizabeth* and *Revenge* classes of battleship and the *Renown* and *Courageous* classes of battlecruisers had only entered service between 1915 and 1917), it was unlikely – and both the Admiralty and industry were aware of this – that there would have been a large-scale replacement programme before the 1930s anyway. Still, the clause also denied a solution to the propping up of industry. Thus, despite the Admiralty's protestations over the policy's likely effect on private yards, the treaty was swiftly ratified.[41] Taken together, these were the catalysts of what would become a radically different approach to the problems the Admiralty and industry collectively faced.

Impact on Industry: Post-Treaty Construction, 1922–1931

The Washington Treaty was a disaster for the private naval arms industry. Although it cannot be said with certainty what might have been built in the absence of a treaty in the late 1920s, what *was* built in the following decade was minuscule by previous standards. When measured in terms of the large, expensive (and most profitable) orders for battleships, battlecruisers or aircraft carriers, just two major ships were laid down (at Armstrong's and Cammell Laird, respectively) between the signing of the treaty and the Japanese invasion of Manchuria five months short of a decade later.[42] These were HMS *Nelson* and HMS *Rodney,* both of which were limited to 35,000 tons – 11,000 tons lighter than the older HMS *Hood* – by the restrictions agreed. While construction of the "medium-sized" (light or heavy cruisers) was initially left unrestricted by

---

[38] Bell, *Royal Navy,* 13.

[39] *Ibid.*

[40] *Ibid.*

[41] S.W. Roskill, *Naval Policy between the Wars: The Period of Anglo-American Antagonism, 1919–1929* (2 vols., London, 1968), II, 22.

[42] Scott, *Vickers,* 216.

the treaty, providing they were no more than 10,000 tons in displacement, this loophole was soon closed and only nineteen were built during between 1922 and 1931. These all ranged in size downwards to 6000 tons and cost between £1.25m and £1.75m, a fifth the cost of the *Hood.* Of this number, more than half were built by the Royal Dockyards, leaving just nine orders for the entirety of private manufacture. John Brown and Fairfield received two each, while Vickers, Palmer, Cammell Laird, Beardmore and Hawthorn Leslie received one, and the rest got nothing at all.[43]

The private industry, comprised of little more than a dozen separate firms until 1930, was otherwise fed for the rest of the decade mainly on a diet of destroyers (nineteen), submarines (fifteen, of which twelve were built by Vickers) and small support ships, none more than a twentieth the size or cost of the *Hood.* To put it another way, the *King George V* class of five battleships, ordered in the space of twelve months under the rearmament programme which began in 1936, was alone worth almost as much to the private naval arms industry as the total income from the entire naval programme between February 1922 and September 1931.[44] The comparison with war conditions is even less favourable; the main class of light cruiser – the "C" class – comprised twenty-eight ships (twenty-one privately manufactured), while the last destroyer scheme before 1918 led to so many orders for the private industry (sixty-seven) that six had to subsequently be transferred to the Royal Dockyards.[45] In all, Britain launched around 300,000 tons of naval vessels between 1920 and the beginning of rearmament sixteen years later, an average output of less than 20,000 tons per year. By comparison, in the five years before the First World War the figure was seven times higher, at 135,000.[46]

Industry Responses, 1922–1926: Price-Cutting, Foreign Orders and Diversification

Faced with the prospect of a long period with limited armament orders, the response of the naval arms manufacturers was two-fold. The first was a collective attempt to undercut rivals, and the second was to diversify into other sectors or other activities in the shipbuilding sector. In normal circumstances, price cuts could help to stimulate demand and offer a path back to profits. In the 1920s, though, this was not the case. The unique circumstances – shipping lines already had a young fleet of post-war ships – meant that cutting prices (a typical cargo

43 McMurtie (ed), *Jane's Fighting Ships*, 202.

44 For a fuller breakdown of costs, see Peebles, *Warshipbuilding*, 105–117.

45 McMurtie (ed), *Jane's Fighting Ships,* 202, 211.

46 Derived from figures in Slaven, *British Shipbuilding,* 63, 73. Indeed, in the ten years before that (1899–1908), it was still more than four times higher, at 82,000.

ship costing over £250,000 in 1920 was on offer for under £60,000 by 1926) only further hurt the firms competing for the few available contracts without attracting any new orders.[47]

Another response was to attempt to sell more vessels to foreign navies. This had the advantage of utilising idle plant and skills in a better way than could be achieved through merchant orders. Not surprisingly, many people saw some merit in this scheme.[48] But while foreign contracts were a part of 1920s (and 1930s) business that should not be overlooked, they did not form a consistent or reliable line of work. Every large naval power possessed its own and often well-developed industry, and those without a large industry did not usually possess a large fleet, and so were not in the market to place massive orders with British firms. Moreover, despite the British belief in the superiority of the domestic industry, such foreign customers had the luxury of approaching experienced constructors on both sides of the Atlantic, as well as in the Far East. The effects of the Washington Treaty were also felt here, for although some navies had some space before reaching their construction "cap," the British industry was competing with the rest of the world for orders. Thus, the reliance was overwhelmingly upon the Admiralty for contracts, which by extension, put it at the behest of its paymaster, the British Treasury.

For its part, the Treasury was either naïve or overly optimistic in its belief that foreign orders would make up for some of what was lost in cancelled British contracts. It also blundered over returning to the Gold Standard in 1925. The belief in running budget surpluses to reduce public debt after the war, and to shore-up faith in the credibility of London as a financial centre, took Britain down the disastrous path towards re-establishing the translation of sterling into gold at a fixed rate. In doing so, it restricted the supply of paper money, since that money had to be backed by gold reserves, which in theory should have had a deflationary effect and held down prices, making British goods more attractive abroad. But by setting the exchange rate at the pre-war level with the US dollar. rather than at the current rate, the effect was to boost the value of the pound which made British goods – including ships – far more expensive to foreign customers, forcing British shipbuilders to further lower prices to stimulate demand.[49] In short, the plan backfired and was scrapped six years later, but only after serious damage had been done to exports.

This was a hurdle that not only affected naval shipbuilding; passenger and merchant shipbuilding had similar problems of chronic overcapacity and severely limited demand. Some firms, however, had pre-war experience in other sectors, allowing for the possibility of these avenues to be reopened with products that

---

47 *Ibid.*, 72, 103.

48 Bell, *Royal Navy,* 159–161.

49 D. Landes, *The Unbound Prometheus: Technological Change and Industrial Development in Western Europe from 1750 to the Present* (Cambridge, 1969), 364.

could be sold in a greater volume in the domestic market. William Beardmore of Dalmuir was perhaps the most active in this regard: by 1925 the firm made cars, motorcycles, diesel engines, locomotives and even aircraft.[50] Armstrong Whitworth – with some pre-war experience in cars and aircraft – was not far behind, adding locomotives to its portfolio in 1920. The same was true for Vickers. Cammell Laird, which only had significant pre-war experience in manufacturing railway wheels, had by 1925 expanded its portfolio to make entire carriages, with some notable success.[51] Others were slower to adjust or failed to adapt at all: Brown, Denny and Fairfield are among a number that stuck rather more rigidly to the manufacture of ships.

Diversification was not, however, a panacea for the structural problems of the naval arms manufacturers. Having the experience and plant capable of manufacturing locomotives or diesel engines gave firms some natural advantages, but it took much more to turn around the fortunes of an entire company. Even with a stream of orders for locomotives, aircraft and some merchant shipbuilding work after 1922, Beardmore was losing significant sums of money by 1924. Locomotives or engines were undoubtedly useful stop-gaps to bolster other income or as additions to shipbuilding and engineering firms in times of need, but these were ultimately still shipbuilding and engineering firms first and foremost, and were large entities that relied on the largest and most expensive orders – naval vessels – to keep their engine plant, slipways, forges and research arms at a minimum level of activity.[52]

Without these orders, diversification was unable to stem the bloodletting, and Beardmore veered toward collapse, which finally arrived in 1930. Armstrong-Whitworth fared only marginally better. After some years of poor results, it was involved in an unequal merger with the largest armaments firm of its kind, Vickers, in 1926.[53] Cammell Laird's rolling stock business was at least moderately successful, and it merged this part of its enterprise with Vickers equivalent to form the new Metropolitan Cammell Carriage Company in 1929. However, even the profits and shared costs realised here were not enough to turn around the fortunes of its parent company entirely.[54]

---

50 See Hume and Moss, *Beardmore*, 180–198, for an overview.

51 Warren, *Steel, Ships and Men,* 210–255.

52 Hume and Moss, *Beardmore*, 181–183.

53 Scott, *Vickers,* 166.

54 Warren, *Steel, Ships and Men,* 215.

Admiralty Responses I: Renewed Treasury Battles, 1924–1926

The inescapable conclusion is that the naval arms industry suffered badly in the 1920s. It was ill-prepared and ill-equipped to diversify production to other sectors, had little recent experience in much except wartime defence work, and owned over-expanded and expensive plant in peacetime. The Treasury's wish to cut expenditure overruled the Admiralty's desire to preserve industrial capacity, but this was only part of a larger narrative of Treasury attempts to curb the power and influence of the Admiralty with regards to spending. This story did not start or finish at Washington; it was only the first blow. Treasury dominance grew even more pronounced throughout the decade and continued until the middle of the 1930s.

It is reasonable to ask if the Admiralty could have more effectively utilised its still considerable power to better effect. The "normal" process in peacetime was for the Admiralty to negotiate directly with the Treasury over the allocation of funds and for a great deal of haggling to go on before the final figure was settled – normally lower than the Admiralty's first "estimates" of needs but higher than the first figure proposed by the Treasury.[55] In the event of an impasse, the First Lord of the Admiralty held the "constitutional right to take disputes with the Chancellor of the Exchequer to the Cabinet," although here he was just one of many men – and below the Chancellor in the pecking order.[56] Thus, a complete breakdown in discussions to the extent that Cabinet involvement was necessary was rare, only occurring over absolutely fundamental differences. This system had worked in the pre-war years, and it was expected to continue to work to settle disputes between the Admiralty and Treasury.

Evidence of the primacy of the Treasury in this period is not hard to find. The Admiralty tried two public schemes to assist private manufacturers, with the aim of providing a steady stream of work within the terms of the Washington Treaty. Both were heavily modified and eventually cancelled by the Treasury. The first was to provide a support fund of £5m through an increase to the navy estimates to provide work and offset unemployment. This was proposed in 1924, although the Treasury cut two-thirds from this figure and refused to renew it in later years.[57] An attempt to guarantee the loans made to shipyards to allow them to undertake contracts before payment had been received lasted only a little longer, and was terminated in 1926.[58]

---

[55] Shay, *British Rearmament,* 25.

[56] TNA, T160/580/10, "Difference between Treasury Control over Civil Departments and over Service Departments," 1934; Gordon, *British Sea Power,* 96–97.

[57] TNA, ADM 167/69, "Minutes of Meeting of the Admiralty Board," paper 1813, 18 February 1924.

[58] Gordon, *British Sea Power,* 77.

Treasury dominance was perhaps most aptly illustrated during the "cruiser crisis" of 1925, which involved the Chancellor of the Exchequer, Winston Churchill, himself a former First Lord of the Admiralty and Minister of Munitions. The crisis unfolded because the Admiralty had recognised the treaty loophole which left cruiser construction limited in size of vessel but unlimited in number. Thus, the Admiralty sought to compensate for the cancellations in large battlecruisers by instead ordering more light cruisers. These were argued for as being essential for protecting the trade routes that Britain, as an island nation with a large overseas Empire, relied on to a much greater degree than other powers.[59]

The existing cruisers, based on pre-war technology which had since rapidly evolved, were now comparatively slow, lacked operating range and were poorly armoured compared to the latest American and Japanese designs. In this sense, they offered the strongest case for modernisation. Moreover, with Japan perceived as a potential future threat, the Royal Navy's senior figures wanted an increased presence in the Far East, and so the Admiralty had announced a new construction programme in 1924. The scheme was intended to last for at least a decade, and would, if fulfilled, have been an important boon to the private industry. The calculation assumed at least seventy cruisers (split between trade protection and supporting the main fleet) to reach "adequate" levels, representing a numerical increase of nineteen vessels above current levels. When including ships in service that were due to reach the end of their useful lives, the conclusion was that at least forty-six new cruisers would be needed within ten years. Based on these plans, the Treasury estimated a budget increase of nearly fifty percent by 1928 on the 1924 naval expenditure estimates, working out at another £25,000,000 annually on top of the £55,000,000 agreed for 1925.[60]

Most of this increase was to be spent on new construction. The wrangling over financing commenced almost immediately. On first appearances, it might have been suggested that Churchill was the ideal Chancellor for the Admiralty to deal with. Quite apart from his previous role as First Lord of the Admiralty, he had been a strong backer of naval expenditure programmes between 1921 and as recently as 1924 was "deeply impressed" by the Admiralty's view of a Far Eastern menace. Moreover, he had under the short Labour government between January and October 1924 attacked its policy of halting work on the British naval base at Singapore.[61] But following the re-election of a Conservative government and his subsequent switch to the task of balancing the budget and tackling a weak economy as Chancellor, his enthusiasm for naval expansion had quickly waned.[62]

Churchill now claimed that the prospects of war with Japan in his lifetime were remote and likely to remain that way, with the implication that the

---

59 Bell, *Churchill and Seapower,* 103–104.

60 *Ibid.*, 103–105.

61 *Ibid.*, 100.

62 J. Neidpath, *The Singapore Naval Base and the Defence of Britain's Eastern Empire, 1919–1941* (Oxford, 1981), 65.

international situation did not resemble that between Britain and Germany before the Great War. Secondly, he challenged the Admiralty's core belief that the improved fleet could land a decisive victory over the Japanese in the event of any war. Britain, he argued, lacked the resources to conduct a prolonged campaign so far from home waters, and the cruiser programme would make little difference to this predicament. Thus, he was in favour instead of a far more limited (and much cheaper) ten-year approach of holding and fortifying the Singapore base, and the commitment to a policy of containment against Japanese intentions in the area.[63]

The Admiralty unsurprisingly disagreed, but Churchill had the ear of a largely sympathetic Stanley Baldwin, the Prime Minister. He continued to argue for reduced expenditure along the lines that there would be little prospect of a war against *any* major naval power for the next twenty years, in effect doubling the time any construction programme would need to be completed, or stopping it altogether until the economy improved.[64] He also took his views to the CID, where he gained support from its Foreign Office representatives. His claim was that a complete cancellation of everything other than submarines and support vessels could save almost £10m a year for two years and would still represent an "ample" increase on the earlier quoted figure of £55m.[65] While support within the CID was good news for Churchill, on its own it was not enough. He still needed Cabinet support for these proposals. In doing so, he was effectively rejecting out of hand the Admiralty view of what would be "necessary" for defence and was taking a substantial risk.

Nevertheless, all but one member of Cabinet sided with Churchill. Only Lord Cecil agreed with the Admiralty (and only then because the First Lord was an "old friend").[66] Given the state of Britain's finances and the proposed spending cuts elsewhere, this was not an unforeseen result. It did not prevent the crisis which then followed. Several senior naval figures responded to the Cabinet's decision by threatening to resign from the CID, the Admiralty and the government if Churchill's restrictions were implemented. Baldwin, under pressure and finding himself in the middle of the wide chasm of opinion that separated the sides, sought a form of compromise – or at least the forestalling of a conclusion – by instructing the CID to form an independent committee to advise on the subject.[67] The deliberations that followed were lengthy, acrimonious, and outside the remit of this work. It does, however, suffice to note three things about them. The first is that they highlight further deterioration in the fractious Admiralty-Treasury relationship following the 1921 expenditure debate and 1924 subsidy; the second is that they represent a rare incidence before Manchuria in 1931 where the CID's subcommittees were called upon by Cabinet to answer

63 *Ibid.*, 104; and Bell, *Churchill and Seapower,* 105.

64 TNA, T161/243/S25613/ANNEX/5, Treasury Supply Papers, Churchill to Baldwin, 15 December 1924.

65 TNA, CAB24/171, Navy Estimates, 29 January 1925.

66 Bell, *Churchill and Seapower,* 107.

67 *Ibid.,* 108.

any questions with immediate consequences; and the third is that the tactic of forming committees to examine and advise (and forestall) was one which would be returned to time and time again throughout the 1930s.

Importantly, the PSOC was not involved in the cruiser debate, for it remained until 1927 under the auspices of the Board of Trade. Thus, the pre-1927 CID was almost wholly basing its decisions on the assessment of threats given by the CoS, and did not approach the question with the impact on industry in mind. Even when chaired by the pro-Navy Cabinet secretary Maurice Hankey, the CID thus still found it difficult to justify the extra expenditure the Admiralty demanded based on the arguments put forward, and eventually also sided with the Cabinet and Treasury. Its conclusion was that the naval budget should be fixed at £58m, which was in fact marginally lower than Churchill's earlier offer.[68]

The report came to the Cabinet in July 1925, and the same pattern was repeated: while the plan to restrict naval expenditure found the support of a sizeable majority, the Admiralty could at least count on Baldwin's wish to avert another round of walkouts and the possibility of resignations. Thus, a compromise was reached by agreeing to a total of seven cruisers being laid down over the next two years, with expenditure on them notionally to be clawed back from future naval estimates and on the condition that the Admiralty agreed to find "offsetting economies" – in other words, to reduce administrative and overhead expenses.[69] This represented a modest improvement in the number of ships built, although it was still six fewer than the Admiralty had originally asked for in the first round of construction alone. All in all, this episode, while neither a complete victory for the Treasury nor a complete humiliation for the Admiralty, demonstrated quite clearly the power of the former's scrutiny over the naval budget.

Like Washington, the cruiser crisis was bad news for private manufacture. Even the most positive development of the talks – the agreement on orders for seven additional ships – had drawbacks. The commitment to reduce Admiralty expenditure in other ways meant that the State-maintained Royal Dockyards (which came with overheads that cost the Admiralty money whether they were actively in use or not) were now under more pressure to pay their own way. The Dockyards were, of course, another centre of capacity, skills and research and as such could never be abolished entirely – even if the Admiralty desired the preservation of private capacity. Thus, it made financial sense to keep Royal Dockyard capacity at least partially occupied, as staffing and overhead costs there were allocated to the Admiralty budget. Therefore, despite the earlier warnings about damage to skills and facilities, the decision was taken that just three of the new ship contracts – for the *London* and *Norfolk* classes – were to be placed with private arms manufacturers.[70]

---

68 TNA, CAB2/4, CID, 199th meeting, 2 April 1925.

69 Bell, *Churchill and Seapower*, 109.

70 McMurtie (ed), *Jane's Fighting Ships*, 38–39.

Chapter Four

From Competition to Collaboration: The Warshipbuilders' Committee and the National Shipbuilders' Security Scheme, 1926–1931

Naval arms manufacturers were in a depressing position at the beginning of 1926, and this suffering prompted the radical action that followed. It was against the backdrop of the cruiser crisis and the naval treaties in the mid-1920s that the WSBC was formed. The committee grew out of other unsuccessful attempts to assist private industry in 1925 and 1926 which drove shipbuilding firms towards collaboration rather than competition. Unable to count on the Admiralty's ability to win increases from the Treasury or Cabinet in the medium term, the individual yards soon realised that the process of competitive tendering was unsustainable and could even drive most of them out of business.

By 1926, it was time for action. The situation was becoming acute: the Coventry Ordnance Works had closed altogether, while Beardmore, Scott and Yarrow had completely run out of profitable work, naval or otherwise, and Palmer and others had fared little better.[1] Sir Alexander Kennedy, the Chairman of Fairfield, summed up the mood of firms in a similar position to his own when he noted despondently that "today private firms [find] themselves burdened with resources and equipment capable of meeting naval requirements far beyond any programme that might for some years to come – if not for ever – likely to be laid down."[2] His words could have come as easily from the First Lord, who after all was providing the hard evidence which supported Beatty's and D'Eyncourt's worst fears.

[1] Peebles, *Warshipbuilding*, 111–112.

[2] Sir Alexander Kennedy, speech at launch of *Taranaki*, *Glasgow Herald*, 12 December 1926.

Admiralty Responses II:
Bending Rules

The Admiralty for its part had long been a champion of state funding to provide a minimum level of orders to ensure firms' survival, even if this would still be a long way from providing an opportunity for the private industry to maintain the world primacy it had enjoyed before Washington. On top of the heated discussions with the Treasury, the Admiralty had also devised several schemes to preserve capacity of key items where few alternative sources of supply existed, but these plans were often poorly conceived and for the most part did not work well.

It has been asserted elsewhere that the Admiralty's concerns about naval gun capacity forced them to give Vickers a "virtual monopoly of contracts" for the rest of the 1920s.[3] This is not strictly true. Guns were a key production bottleneck, but it is important to distinguish between gun mountings – the large, swivelling, turret part of the gun – and the other parts of the gun, including the forging (the tubes and the barrel) and surrounding mechanism. In the case of mountings, the Admiralty could hardly *give* Vickers a monopoly, since from 1925 it was the only firm left in the country capable of making them, so it had a monopoly by default. In terms of forgings, the Admiralty split the small amount of available work proportionally among Vickers, Armstrong's and Beardmore.[4] Thus, although the Admiralty was very worried about the condition of private facilities, the step was not perhaps as radical as it might seem.

The gun case, however, brings us to a tactic that was tried and re-tried throughout the second half of the 1920s – finding ways to bend normal contract procedures to ensure the release of funds to firms in a more predictable fashion. The key figure here was the Third Sea Lord and Controller of the Navy from 1925 to 1928, Sir Ernle Chatfield. Chatfield, who sat on the PSOC and was the direct superior of the Engineer-in-Chief of the Fleet (at this point Harold Brown, Supply Committee III's chairman), had overall responsibility for procurement and had expressed his concern at the decline in naval capacity, noting the "[great] importance of retaining latent strength for some future danger." Because of this, and the responsibility he felt "for the future material development of the Navy," Chatfield believed he had to look to "the great private firms."[5]

Even before the cruiser crisis, Chatfield held discussions with steel manufacturers capable of making armour plate – another major bottleneck – for an additional cost to be added to bids to cover "dead charges for plant not fully used but kept in being."[6] After the failure to get the funds for direct assistance from

[3] M.M. Postan, *British War Production* (London, 1952), 48.

[4] Gordon, *British Sea Power,* 86–87; and Buxton and Johnston, *Battleship Builders,* 174.

[5] E. Chatfield, *It Might Happen Again: The Navy and Defence* (London, 1947), 38.

[6] Gordon, *British Sea Power,* 82.

the Treasury in 1924, Chatfield spearheaded an agreement in 1925 with firms to pay a higher price for small orders of armour – in other words, to provide a subsidy in the form of a cost premium to steel firms to keep otherwise idle plant in operation. He went to great lengths to keep this arrangement secret, almost certainly because of the failure of the two earlier schemes and the cruiser crisis, so he only informed the PSOC about it in guarded terms.[7] Indeed, when the Auditor-General found out about it the following year, the Treasury immediately protested; after a payment of £60,000 was made despite no armour being ordered in 1929, the scheme collapsed.[8]

Compared with the attempt to get £5m added to the naval budget, this scheme represented only a limited attempt at assistance, but it nevertheless still failed. This forced a change of tack, pushing the Admiralty towards working with industry to form a rota to share orders and ensure the steadier and more predictable stream of funds which had formed the basis of earlier schemes. This is what would become the WSBC, an extremely important organisation in the story of 1920s and 1930s armament manufacture, which grew into a sophisticated price-fixing cartel during the rearmament era. Very little has been written about the operation of this organisation, and extensive documentation is almost non-existent. Indeed, what does survive can be found only in the records of two shipbuilders – Scott and Brown. While both collections contain letters and circulars from the other yards, such as Vickers, the other side of the paper trail does not appear in the corresponding Vickers archive. This gives rise to the possibility that they may have been deliberately destroyed. Moreover, even in the two collections where material on the WSBC still exists, it is either fragmentary or extensively fire damaged. Given the conduct of the committee, there are many reasons for why the committee might have wished to destroy its records. Fortunately, enough remains to allow at least an outline of the committee's early work and thus the construction of a coherent narrative.[9]

---

[7] TNA, CAB60/2, 24th meeting of the PSOC, 25 February 1926.

[8] Gordon, *British Sea Power,* 84.

[9] The main bulk documentary evidence for the WSBC comes from four badly fire-damaged boxes in Scott's of Greenock shipbuilder's records at the Scottish Business Archives, University of Glasgow. Correspondence not only with Barr at the SEF but also with most other committee firms, along with minutes and circulated notes, survives in reasonable quantities. From this, it is possible to establish the prominent role several company directors played; yet Vickers, like most other yards involved in regular correspondence, have no traces in their own collections. It is likely therefore that the intention was to destroy all evidence of their dealings once the Admiralty instigated cost investigations in 1941. Most of what follows is drawn from the fire-damaged letter folders in GUAS, GD319 12/7/5–7.

Composition, Organisation and Membership: The Destroyer Rota, 1926–1929

On the infrequent occasions when the WSBC is mentioned in broader published studies, some useful, though often contradictory, points have emerged. Though only a handful of historians have even mentioned the organisation's existence, they generally agree on three points: that it began as an unofficial and secretive "rota" designed to share Admiralty work on destroyers; that it had no relation to the organisation of the same name which existed as an employers' union from 1899 to 1911;[10] and that the Admiralty either directly assisted the group's creation or at the very least was willing to overlook its existence for its own benefit as well as for that of the industry as a whole.[11] On the last point, however, there is some difference of opinion, although given the nature and scarcity of sources (and that the WBSC is usually of tangential interest to other studies) this is not surprising. For instance, Hugh Peebles and Anthony Slaven conclude separately that the Admiralty assisted warship builders with the operation of an informal destroyer rota in the 1920s due to an arrangement dating from November 1926,[12] while Lewis Johnman and Hugh Murphy believe that the Admiralty and government had no knowledge whatsoever of this "price-fixing ring" until the group was forced to admit its existence in front of the Royal Commission for the Private Manufacture of Arms in 1935. Johnman and Murphy also contend that the Admiralty was then prepared to deal with the WSBC as a single body.[13] The shipyard sources, in combination with Admiralty and other papers, shine new light on the WSBC and its motives; help to show how much was (and frequently was not) known about its existence; and add colour to what has so far been only a brief discussion of the highly secretive nature of its business.

In its initial form in 1926, the WSBC was composed of fifteen firms – the same as in 1918, minus the Coventry Ordnance Works, which had folded in 1925.[14] In addition, Vickers and Armstrong-Whitworth merged on 1 January 1927. Beardmore had ceased shipbuilding by 1930, followed by Palmers in 1933, while Harland & Wolff at Belfast was added in November 1930. Otherwise, the membership remained constant until at least 1942.[15] Indeed, excluding the government-owned Royal Dockyards, no naval vessel carrying a gun larger than

[10] For more details on this group, see University of Warwick, Modern Records Centre, MSS 237/6/A, Federated Admiralty Contractors papers.

[11] L. Johnman and H. Murphy, "Welding and the Shipbuilding Industry," in R. Harding (ed.), *The Royal Navy, 1930–2000* (London, 2005), 101; Slaven, "A Shipyard in Depression," 200; and Peebles, *Warshipbuilding,* 111.

[12] Slaven, "A Shipyard in Depression," 200; Peebles, *Warshipbuilding,* 111.

[13] Johnman and Murphy, "Welding," 101.

[14] GUAS, GD319 12/7/7, table 4.2, "Shipbuilder Codenames as of 1932 *SECRET: Minutes of Private Arrangement made by letter concerning Minespools 32.*"

[15] GUAS, GD319 12/7/6.

six inches or weighing more than 1000 tons was built for any navy in the world in British yards at any point between the wars by any firm except those listed. On occasion, some specialist engine manufacturers, like Parsons and Wallsend Slipway, were included in discussions, although not in the committee's first years when it operated the destroyer rota only. These firms were, not surprisingly, only part of discussions for contracts or bids for machinery – for example, the engines for HMS *Warspite* – in which Parsons was the successful "bidder."

The WSBC was created a time of broader industry collaboration. The Shipbuilding Conference, formed in 1928, was a parallel trade organisation born out of the desire to deal with the "rather uncoordinated state of the industry on the commercial and economic side ... which in the grievous economic circumstances ... both existing and imminent, threatened grave danger to the ... shipbuilding industry as a whole."[16] The Conference was a much broader initiative which, in addition to all the WSBC members, brought together dozens of shipbuilders of all kinds and instigated an initiative for an industry-wide "tendering expenses scheme" to pay a one-percent levy on successful tenders into a collective pot to reimburse unsuccessful bidders to avoid firms lowering their estimates and profits margins to secure contracts.[17]

The WSBC was very powerful, even in the 1920s when individual firms were at their weakest. After all, no other firms in Britain, except those in the group, could meet the demands of constructing a complex warship or had any experience of Admiralty work. Furthermore, it was widely accepted that it was politically impossible – given the precarious position but also the world-leading reputation of British shipbuilding and the degree of technical secrecy involved – for the Admiralty to place an order abroad, even if it was technically possible or financially more prudent. Instead, the aim in the 1920s was to assist British firms to win orders from foreign navies.[18] This sits in stark contrast to the Shipbuilding Conference, which did face competition from abroad and had a wide range of different customers. Thus, even if the WSBC was operating as a subset of members of the Conference, it had some very notable advantages over it.[19]

The key to comprehend the WSBC's formation into a work-sharing rota is to understand not only the shared interests of the individual members but also the shared concern of the Admiralty and the industry to preserve industrial capacity. The Admiralty, like the industry, had reiterated its fear of a "race to the bottom"

---

16 J.R. Parkinson, *The Economics of Shipbuilding in the United Kingdom* (Cambridge, 1960), 136.

17 NMM, SRNA4/S50, "Proposals for a Tendering Expenses Scheme and Other Methods to Aid Shipbuilding," 1928.

18 Bell, *Royal Navy,* 149–157.

19 H. Lyon, "The Relations between the Admiralty and Private Industry in the Development of Warships," in B. Ranft (ed.), *Technical Change and British Naval Policy* (London, 1977), 62–64. Lyon mentions the Shipbuilding Conference but does not appear to know about the WSBC.

if strictly secret competitive tendering continued, a practice which could drive yards out of business as each fought to win contracts at unsustainable prices.[20] From the Admiralty's perspective, such cost-cutting had obvious disadvantages. While it might temporarily provide better value for money, only the most efficient yards would survive, and thus total capacity among specialist yards would be severely restricted in future times of need. Moreover, upon a return to relative prosperity, when truly competitive tendering could be reinstated, the more firms still in existence the better to ensure the best value for money and that facilities remained up to date.

Despite this, not every yard was initially convinced of the merits of the group. For example, Eustace Tennyson D'Eyncourt, the former Director of Naval Construction (DNC, and not to be confused with the Director of Navy Contracts), at this point a director of Armstrong's (a position he held until 1927), expressed reservations about agreeing prices and implementing a levy. At a meeting in June 1926 he reported unequivocally that his "Chairman would have nothing to do" with such schemes, although it is not clear what his own personal views were, especially considering his earlier warnings about the decline of the naval armaments industry.[21] Nevertheless, it seems that his chairman was quickly talked around, as the firm was listed only a few months later as a founding participant.

The first operation of the scheme, agreed at a meeting that D'Eyncourt attended, showed the clear advantages for both the shipyards and the Admiralty. The firms first devised a three-year rota relating only to destroyer orders. A very brief attempt had been made earlier (in 1925) to devise a cruiser rota, but the cruiser crisis killed the scheme off and shifted minds elsewhere by 1926. Each firm was allocated a position in the rota simply by the drawing of lots and could bid at a price agreed among the members, a level at which a small profit was anticipated, thus allowing a more even spread of contracts to the existing yards and avoiding a race to the bottom.[22]

In the circumstances, it could be argued that this was not highly controversial – while it certainly broke tendering rules, it carried a reasonably low risk to the taxpayer or the Admiralty. After all, destroyers were constructed by all the member firms to a standard design, were comparatively small and inexpensive, were unrestricted by the Washington Treaty, and were ordered relatively frequently. At the same time, the Admiralty retained the right to inspect the financial records of all firms and had the ability to compare costs against previous orders or for ships constructed in the Royal Dockyards.[23] Furthermore, with over a dozen competitors, any excessive bids in the scheme's early phases

20 TNA, CAB60/2, 24th meeting of the PSOC, 25 February 1926.

21 GUAS, GD319/12/7/6, Brown to Scott, 17 June 1926.

22 GUAS, GD319/12/7/6, "Meeting at Hotel Cecil – Hull and Machinery for Admiralty Contracts," 19 December 1926.

23 Peden, *British Rearmament and the Treasury*, 50.

obviously carried the risk that the firm in question might not be invited to tender in the future. Thus, on the face of it, a system of checks and balances enabled the rota to work while keeping prices from escalating out of control. In sum, given the troubled picture for shipbuilding in the mid-1920s, and the concern for the future existence of some of the Admiralty's major partners, the operation of a rota for destroyers seemed to offer some security in a climate of scarce orders, while also enabling the private industry to keep facilities in use and skills sharp in case of future need.

The question of how much was known about the WSBC by the Admiralty does not, however, have an entirely clear answer. This is perhaps to be expected – the Admiralty certainly had an interest in at least keeping up the pretence that it knew nothing. Moreover, in practice the rota did not require direct Admiralty involvement; firms in the group could, in theory, have bid at an agreed price without any outside intervention. So, while Peebles and Slaven both explicitly state that the rota was drawn up with Admiralty assistance (and Murphy and Johnman believe that the Admiralty had no knowledge of it), it is difficult from the surviving documents to pin down exactly what form any assistance might have taken. On one hand, it *cannot* be said with certainty that the Admiralty had full knowledge of the 1926 agreement, or more specifically, how the firms communicated with each other. There is no document which unequivocally proves either. Indeed, the Admiralty never officially admitted knowing anything about the WSBC, even in testimony before the 1936 Royal Commission.[24]

On the other hand, there is evidence which suggests that tacit Admiralty approval was given to the operation of this ring and its one-percent levy. Firstly, there is the precedent which had previously been set by the Admiralty to assist firms directly. While this had started as early as 1920, there had been the more recent cases of the short-lived armour subsidy and the increase in naval estimates to try to alleviate unemployment. In each case, the Admiralty failed to get the Treasury to agree to provide sufficient funds to maintain idle plant. On the armour subsidy, the firms that formed the agreement included Cammell Laird, Vickers, Armstrong-Whitworth and Beardmore, along with other exclusively steelmaking firms. In addition, D'Eyncourt had a close professional relationship with Chatfield since before the Washington Treaty.[25] Thus, it is nearly certain that most, if not all, WSBC members knew of the previous subsidy and assistance schemes, as well as Chatfield's desire to assist private builders. Therefore, it is also reasonable to assume that Chatfield had a hand in organising a destroyer rota. As this was perhaps even more contentious than the price agreement for armour, it is not surprising that the Admiralty never admitted knowledge of it. Nor should it be surprising that very few other references remain of these agreements in the papers of the WSBC members or in Chatfield's diaries, letters or memoirs.

---

24 Murphy and Johnman, "Welding," 101–102.

25 D'Eyncourt, *A Shipbuilder's Yarn,* 140–141.

This conclusion – broadly in line with Peebles and Slaven – does not, however, assume that the arrangement remained this way. There is far less evidence to suggest that the Admiralty or Chatfield's successor as Controller (Sir Roger Backhouse) knew that the group continued to exist beyond the initial "emergency" period between 1926 and the end of the initial three-year rota in 1929, by which point the Shipbuilding Conference and an industry-wide levy had been established (and the Conference was a much higher profile organisation that could reasonably have been believed to have superseded any rota). Nor does it appear that the Admiralty had any idea of how the group worked in practice or how it transformed from its early days into a powerful rearmament cartel. Despite the sharing of staff between industry and the Admiralty, there is no evidence that information about the increased scope of the WSBC's later schemes was understood at all in Whitehall. In other words, it was a one-way relationship after the termination of the destroyer rota. From later evidence during the rearmament era, when the WSBC desperately tried to stop the Admiralty from uncovering the extent of its members' profiteering, it can be reasonably concluded that, as Murphy and Johnman suggest, the Admiralty knew almost nothing at all about the inner workings of, the Warshipbuilders Committee until the evidence presented to the Royal Commission on the Private Manufacture of Armaments in 1935, by which point Chatfield was First Sea Lord. Yet even after this date it continued to know embarrassingly little and was continually outmanoeuvred until the middle of the Second World War. In short, the rota drawn up for destroyers in the 1920s can be viewed as a direct successor to earlier Admiralty assistance schemes, while the extensive profiteering which followed in the later 1930s was a different matter altogether.

To return to the organisation and workings of the WSBC, we know that the members met as a group for the first time in London's Hotel Cecil near the headquarters of the Shipbuilding Employers' Federation (SEF) in February 1926. Meetings subsequently took place with a representative from the SEF in attendance until the hotel closed in 1930; thereafter, they took place either in a private room at the headquarters of the SEF at 13 Victoria Street, London, or the nearby Hotel Victoria. The SEF's controller from 1923, former shipbuilder and industrialist John Barr, was the man chosen to be responsible for the lines of communication between the WSBC yards and the Federation, a position he held continuously until 1937. The representatives from the firms themselves were drawn only from the highest echelons; in most cases, it was the chairman, or a suitably senior director. Moreover, a study of the directors' minutes of the WSBC's member firms leads to the conclusion that discussions were kept very secret, for despite the frequent attendance of men like Kennedy (Fairfield) and Craven (Vickers) who chaired their respective boards, there is no mention of the WSBC's existence at their respective company board levels.[26] Indeed, outside

[26] See, for example, GCC, Mitchell Library, UCS2/1/2, Fairfield, directors' minutes;

of the very top ranks of the SEF and the firms involved, few knew even of the committee's existence, let alone the scope of its operations.

Nevertheless, it was also the case that before 1929 the information shared was limited and relatively uncontentious; it remained only a destroyer rota, and "normal" tendering practices continued for the few larger orders that were available, including for cruisers. The rota was not particularly sophisticated; Beardmore, Hawthorn Leslie, Palmers, Fairfield, Cammell Laird and Vickers were the first names drawn from a hat to receive orders.[27] Nor did the profit margin involve much in the way of sophisticated calculation – a round number equivalent to approximately ten percent of the costs was to be added to the final bids.[28] Perhaps because of the rather rudimentary methods of allocating orders and setting prices, the first iteration of scheme was thus also ineffective at solving the problems facing the industry. Profits on the contracts remained low, and in some cases yards still made small losses. Fairfield, for instance, lost £1375 on one order, while Brown at Clydebank lost almost £10,000 on a contract for two destroyers.[29] Still, these were relatively minor problems compared to the benefit of doing away with the cutthroat competition that the industry and the Admiralty feared. These results can also be interpreted as firms adopting a naturally cautious approach to stating their costs and not seeking to inflate prices to levels that would be a clear hike over previous quotes, and thus risking not being invited to tender for orders in the future.

By the start of the 1930s, however, a shift occurred: small but significant profits were being made on almost every naval order – including, but not limited to, destroyers. This was in stark contrast to losses that continued to be incurred on merchant and passenger work. At John Brown, compared with the loss on two "A-Class" destroyers, the two "F-Class" contracts four years later netted an extra £87,000, a marked improvement from the merchant situation from 1922 to 1928, where twenty-three orders in total netted the company a combined loss of £211,000.[30] At Fairfield, the small loss on the naval machinery contract was an exception rather than the rule, compared with twelve out of fifteen merchant orders in the same period that failed to cover costs.[31] Although returns on Admiralty work rarely yielded more than a modest profit before 1931, the WSBC appeared to at least be safeguarding a price "floor" below which tenders would not sink.

This also indicated a broadening of the group's remit from the initial purpose of cooperation on destroyer contracts, which were small and numerous, to those

CUL, MS Vickers, Vickers Armstrong, directors' minutes; and GUAS, UCS1/5/6, John Brown Clydebank, directors' minutes.

27 GUAS, GD319/12/7/6, "Meeting at Hotel Cecil – Hull and Machinery for Admiralty Contracts," 19 December 1926. The rest of the rota has been lost to fire damage.

28 *Ibid.*

29 Peebles, *Warshipbuilding*, 111.

30 Slaven, "A Shipyard in Depression," 198.

31 Peebles, *Warshipbuilding*, 110.

for larger ships. Pinpointing precisely when and how this shift came about is difficult; although evidence from 1930 may help to shed some light. That year was one of transition for the WSBC, following the bottoming-out of new orders in 1929. The initial rota had been concluded with the launch of the last of the "A-class" destroyers in October 1929, and had been extended to the "B" and "C" classes, the latter of which had been allocated in the summer of 1930, with work due to begin that autumn in the respective yards.[32] At this point, the SEF was informed through "unnamed sources" that Harland & Wolff (H&W) of Belfast was also being invited to tender for Admiralty contracts for an upcoming cruiser class, information that was immediately passed on to the members of the WSBC.[33] H&W, although not a frequent supplier to the Admiralty, did have some naval experience during the Great War, with the construction of the cruiser HMS *Glorious,* and also had facilities capable of handling large vessels.[34] At the same time, the WSBC member Beardmore was in the process of voluntarily liquidation under the National Shipbuilding Security (NSS) scheme and was no longer taking part in warship construction. Thus, when Captain T.E. Crease, Barr's deputy, communicated the developments at H&W to James Brown of Scott's, he argued that "this makes a difference, as they [H&W] are not parties to the current arrangement and must be brought in somehow."[35]

Other members agreed. Craven from Vickers and Batey from Hawthorn Leslie joined Crease as "representatives from the group of firms" and met Rebbeck of Harland & Wolff on 12 November 1930 to put across their proposals for joining the rota, where they "emphasised the necessity for discretion and absolute secrecy."[36] H&W agreed to the terms the following day and was formally admitted to the group at a meeting two weeks later. In the next rota, firms agreed that "it was essential that [the] arrangement should be made to ensure the reasonable geographical distribution of work." Thus, the yards were divided into three geographical areas: the Tyne (known as Group No. 1), the Clyde (No. 2) and the North West (No. 3). Firms were only allowed to switch places on the rota with other firms within the same geographical group, or if it was otherwise unanimously agreed. Crucially, the list omitted the White and Thornycroft shipbuilding companies in the South of England. These firms were specialist builders of destroyers and smaller naval vessels, so they were omitted because of the nature of the new scheme, which agreed "that the rota should provide for cruisers, aircraft carriers and all auxiliary warships of a value exceeding £600,000 for hull and machinery only, for the Admiralty and Dominion navies."[37]

---

32 J. English, *Amazon to Ivanhoe: British Standard Destroyers of the 1930s* (Kendal, 1993), 30–33.

33 GUAS, GD319/12/7/6, Crease to Brown, 29 October 1930.

34 Buxton and Johnston, *Battleship Builders,* 84.

35 GUAS, GD319/12/7/6, Crease to Brown, 29 October 1930.

36 *Ibid.*

37 GUAS, GD319/12/7/6, Meeting of Warship Group, December 1930.

The group was clearly no longer a destroyer rota over a fixed time-scale. It was now definitively and consciously developing a much larger scheme. The fact that "secrecy" was mentioned so clearly is also striking. Who would H&W need to be careful of informing? Given that the new agreement was for Admiralty and Dominion contracts, it is reasonable to assume that this was a deliberate attempt to eliminate any outside knowledge, the Admiralty included. After all, the last time the Treasury had found out about an Admiralty assistance scheme – when Chatfield informed the PSOC which in turn informed the Auditor-General – it had led to the collapse of Chatfield's armour subsidy in 1929.[38] Thus, it is particularly noteworthy that the Admiralty took no part in these meetings and that there was no attempt to include it in discussions or notify it of the decision that H&W was to be "brought in." It is thus reasonable to conclude that the Admiralty had no continuing knowledge of the operations of the WSBC.

Moreover, it appears clear that the WSBC's members believed that this work-sharing agreement brought benefits that could not be discarded or disclosed lightly, and that it was therefore critical that the WSBC continue to include every private manufacturer capable of making any kind of naval vessel for the Admiralty. This new plan had to be (and was) more sophisticated, with different rotas and agreements for different kinds of ship. Most importantly, as evinced by the discussions with Harland & Wolff, it had to be secret.

Developing the Lines of Communication after 1929

The financial returns on Admiralty contracts show that while the WSBC's destroyer rota prevented cutthroat competition, it did not guarantee handsome profit margins. Yet viewing the early years before 1931 solely from a profits perspective is to underestimate its significance as an organisation. As one might expect, income and profits in these lean years of defence expenditure and only modest civilian construction were low compared with the demand for ships during the Great War or during rearmament in the second half of the 1930s.[39] The crucial point is that the firms had moved from a climate of competition to an information-sharing forum and a culture of cooperation, while keeping the existence of such cooperation secret. This asymmetry of information after the conclusion of the first destroyer agreement was the WSBC's greatest strength. As a forum for discussion with the power to conclude deals among rival firms, the WSBC possessed the most important tools necessary for manipulating the market to its own benefit: a united membership and a customer (the Admiralty) that could and would buy only from these members. As such, the WSBC represented a radical departure in the history of British naval shipbuilding and arms

38 Gordon, *British Sea Power*, 84.

39 Peebles, *Warshipbuilding*, 124.

contracting by deviating from competitive tendering and replacing the process with agreed prices that led the customer to believe that the lowest bid represented the best value. In practice this fatally undermined the initial benefits envisaged by the Admiralty and explains in part the handsome profits that returned during rearmament.

The enhanced process of cooperation after 1930 worked in the same basic way for more than a decade. Upon receiving a tender from the Director of Navy Contracts (again, not to be confused with the Director of Naval Construction or DNC), each firm wrote to Barr to notify him of the Admiralty's intention to place an order. In a typical example, James Brown at Scott "begged to advise the committee" that his company had "received the usual enquiry for the construction and completion in all respects of Hull and Machinery of one and each of the two destroyers for H.M. Navy," before detailing the exact deadline for Admiralty receipt of the tender and communicating the Admiralty's instruction that the tender was strictly confidential.[40] Barr (or occasionally, Crease), collected such notifications while also writing to firms which had failed to send word to confirm that they had also been invited to tender.[41] Such information sharing allowed Barr to do two things: first, to confirm that the tenders were being invited from all eligible firms, in other words to ensure that the Admiralty was not unduly favouring one yard over another; and second, to level the playing field and alert yards to tenders they had not yet received, but could reasonably expect at a later date, giving them as much time as possible to prepare cost estimates. The driving motivation behind these actions was to circulate intelligence on Admiralty procedures; by ensuring the continuance of tenders being sent to all eligible firms, the committee could in turn continue in the belief that no yard was being excluded on a price or quality basis, and thus could continue to design effective rotas for tenders at prices that the committee considered would allow for acceptable profits.

Barr's next task was to ascertain which firms were interested in "winning" the contract. In the disarmament years, this included practically every yard able to undertake an order, while in the rearmament period it was more common for several firms to withdraw. The same process was followed: firms would notify Barr of their intent to bid, and Barr (and later Crease)[42] would coordinate

---

40 See, for example, GUAS, GD319/12/7/6, Brown to Barr, 13 October 1933 or Brown to Crease, 18 November 1935.

41 GUAS, GD319/12/7/6, Barr to Alexander, 6 December 1933.

42 Murphy writes that Crease was appointed as the Admiralty's "liaison" to the committee after 1936, with the stated intention of keeping the Admiralty in the loop as to what the WSBC were doing in the aftermath of the revelations at the Royal Commission. Despite his position as a former senior naval officer, the fact that he had been working with the committee for several years beforehand suggests either that he was not forthcoming with information to the Admiralty before 1936, or that the "appointment" was merely a continuation of past practice. It is doubtful he was any more willing to divulge details

efforts to get representatives from each to a meeting where the most accurate figures available would be discussed openly with all manufacturers, so that the group knew, to the penny, what each firm believed it would cost to produce the vessel required by the Admiralty.[43] Such information was then compiled into lists and circulated, so that each committee member had a written record of its own costs relative to those of others in the group. In rarer but not necessarily infrequent cases, particularly when meetings had been called at short notice, firms wrote to each other to provide costs. To give one example, in January 1929, Sir Harold E. Yarrow, managing director of the shipbuilders of the same name in Scotstoun, Glasgow, wrote to James Brown at Scott giving him a full and exact breakdown of their estimated costs for a destroyer leader, a destroyer, and a depot ship.[44]

That the men at the top of what should ordinarily have been rival organisations were willing to disclose to their opposite numbers how they arrived at final prices is, in some ways, astonishing. This was well beyond the terms of the initial rota, was risky, and perhaps even undermined "normal" business practice, namely the desire to keep information about costs and overheads secret, lest a rival use it to their advantage. This said, gentleman's agreements of this sort were not uncommon before 1945. The Shipbuilding Conference of 1928 was a variation on a theme: many similar national shipping schemes existed to regulate freight rates or passenger fares.[45] Indeed, many of these were publicly backed by that country's government to protect its own industry.[46] The Board of Trade even noted during the Second World War that "the variety of arrangements is striking, and attests to the ingenuity of industrialists."[47] What set the WSBC apart, however, was that it was a group of firms working against a single customer – the government – to regulate prices. The Shipbuilding Conference agreed an industry-wide levy that covered tendering costs, but its members had no control over its customers – who could freely choose to go elsewhere if the price was not right – nor did its member firms share their own breakdown of costs with each other.

A mix of historical factors and contemporary circumstances made the WSBC unique. The benefits of collaboration, the value of the shared information, and the prospects to negotiate better prices – even if the contracts had to be shared between rivals – outweighed the drawbacks and reflected in part the desperate

afterwards either, given the Admiralty's continued inability to regulate the committee in the years after the commission.

43 GUAS, GD319/12/7/6. "Estimate," undated.

44 GUAS, GD319/12/7/6. Yarrow to Brown, 28 January 1929.

45 W. Sjostrom, "Collusion in Ocean Shipping: A Test of Monopoly and Empty Core Models," *Journal of Political Economy,* 97/5 (1989), 1170–1171.

46 J. Fear, "Cartels," in G. Jones and J. Zeitlin (eds.), *Oxford Handbook of Business History* (Oxford, 2008), 273–277.

47 *Ibid*, 273.

situation for naval shipbuilders in the 1920s and early 1930s. More simply, these episodes highlight the immense collaboration between all member firms.

The one-percent "tendering expenses" levy added to the final cost was therefore not the only good reason for keeping the WSBC together. In addition to the access to information, there was the potential for pooled resources, steady allocations of work, and a high price floor for orders. The kick-back that the levy provided to cover the costs of preparing tenders sweetened matters, but taken together all these factors provided a strong disincentive to leave the committee. Indeed, while D'Eyncourt initially reported that his chairman at Armstrong did not like the idea in 1926 (its own weak financial position forced it to merge with Vickers in January 1927, so it was not a member of the group in that guise for long at any rate), any warship firm would have found existence outside of the group significantly more difficult. For without knowing the "winning" price a competitor would have been forced to undercut its rivals, and in not knowing the prices that rivals were bidding, it would likely have had been faced with a return to cutthroat and loss-making bids to gain contracts. It is therefore not surprising that from the WSBC's earliest days until well into the Second World War the membership remained unshakeable.

There was also an external "push" that encouraged this strength-in-numbers approach. This was precipitated in 1929, although it did not manifest itself fully until rearmament. The election of the second MacDonald Labour government in June 1929, and its public commitment to a further curtailment of defence spending, was almost certainly a catalyst that strengthened the bonds among firms. It also underscored the danger of being seen to collude on stabilising prices when the government of the day intended to – with significant public support – pursue disarmament and cut arms spending further. Thus, like the first steps towards cooperation between the yards had been the result of external circumstances, the 1929 election highlighted to each yard the importance of sticking together. Moreover, if the election pledges of the Labour Party – "arbitration and disarmament" – were anything to go by, the years to come appeared in 1929 even worse than those which had just passed.[48] In other words, this was no time to end the relationship. The members of the WSBC, probably even before the admission of H&W, had realised that it was in their collective interest to continue to utilise the lines of communication that had been built up until market conditions substantially improved.

It might appear that one other possible solution was a redoubling of Business-Admiralty unity in much the same way as had occurred in 1926, and to continue lobbying for naval expenditures. But MacDonald's visions for disarmament in 1929, and the prospect of another arms limitation conference in London in 1930, greatly diminished any prospects for such a plan. Thus, as the prospects for any rise

[48] R. Vickers, *The Labour Party and the World: The Evolution of Labour's Foreign Policy, 1900–1951* (Manchester, 2003), 93–94.

in naval spending looked undeniably grim in the second half of 1929 (which proved to be one of the industry's worst years),[49] the response was to expand the rota and ensure that profits were returned on all ships, not just destroyers. On one level, this was rational. It was also illegal, and went well beyond the original scheme.

It has already been noted that the Admiralty retained the right to inspect the financial records of the firms that supplied the navy; in practice, though, it did not take up this option at any point during the years 1929–1940. This oversight, although large, was not necessarily the only reason the WSBC could continue operating in the 1930s, for with the methods outlined above, it is unlikely the ring would have been uncovered unless there was a particularly thorough examination of practically all WSBC members. Moreover, while the CID had convened a "contracts coordinating committee ... to secure economy and eliminate the forcing up of prices by competition" since 1920, it was concerned about precisely the opposite effect, namely a major crisis causing inter-service rivalry for resources and thus industry selling to the highest bidder, and as such was not an effective check on WSBC tenders.[50]

The members of the committee were confident that their financial records would not be examined thoroughly enough to reveal their plans. In this sense the long-standing relationship between the Admiralty and most of the WSBC firms worked in the latter's favour. Indeed, E.C. Jubb, the Director of Navy Contracts, wrote in 1936 that the system of competitive tendering remained "one of the greatest safeguards against undue profits."[51] Whatever the underlying reasons, the WSBC took no chances, and from 1930 it certainly grasped the opportunity to develop a covert organisation that would help keep it free from unwanted attention.

One obvious step that was taken was to remove reference to its members. Therefore, by the beginning of 1931, many (but confusingly, not all) mentions of firms in the minutes of meetings and correspondence were in code. This apparently followed a set pattern: Clyde yard "one" was "CONE," Clyde two was "CETU," and so on, although it was not always this clear. For example, while Stephen, Scott and Denny were on the Clyde, and Swan Hunter and Hawthorn Leslie were on the Tyne, and thus fit the pattern, Yarrow and Fairfield (also on the Clyde) were "D," while yards as far apart as Belfast and the Isle of Wight were grouped under "M." Furthermore, with barely a dozen members, there appear to be spaces left in the code to which no yard in the group was allocated (see Table 4.1, below). Nevertheless, the codes, much like the first attempt to draw a destroyer rota from a hat, seem unsophisticated and amateurish. Handwritten scribbles on several pieces of correspondence put the names of firms next to codes, rendering them

---

49 See Johnman and Murphy, *British Shipbuilding and the State,* table 3. Unemployment in private shipbuilding peaked in October 1929 at over forty percent.

50 N. Rollings, "Whitehall and the Control of Prices and Profits in a Major War, 1919–1939," *Historical Journal* 44/2 (2001), 524.

51 CUL, MS Templewood, part 9, file 5, "Admiralty Contracts, Prevention of Profiteering," 17 July 1936.

Table 4.1:
Shipbuilder Codenames as of 1932

Shipbuilder	Codename	Area
Stephen	CONE	Clyde
Scott	CETU	Clyde
Denny	CESEX	Clyde
John Brown	CESEVN	Clyde
Yarrow	DETU	Clyde
Fairfield	DETRE	Clyde
Thornycroft	MOTRE	Hampshire
White	MOFOR	Isle of Wight
Cammell Laird	MOFIV	Birkenhead
Vickers Armstrong	MOSEX	Tyne
Harland & Wolff	MOTEN	Belfast
Swan Hunter	TONE	Tyne
Hawthorn Leslie	TEATE	Tyne
Palmer	Not listed	Tyne

Source: *SECRET: Minutes of Private Arrangement made by letter concerning "Minespools 32,"* GUAS: GD319 12/7/6

useless, while on other occasions the firms dropped the code names altogether and lapsed into referring to each other in the usual way. Had it not been for the urge for "utmost secrecy" from Barr at the SEF (which indicates that the group knew that what it was doing was likely to land its members in serious trouble), one wonders if these code names were ever intended to fool anyone.

In addition, minutes with sensitive information (including details of prices) were stamped "secret" rather than just "private," often were undated, and were not sent by mail. At least in the cases of the surviving "secret" files, these appear to have been circulated at face-to-face meetings in London, which minimised the odds of interception.[52] Highly ambiguous telegrams – saying merely "pool abandoned" – also exist between firms.[53] In sum, the WSBC committee was acutely aware of the illegality of their actions, if not always particularly good at devising methods to keep those actions concealed.

[52] GUAS, GD319/12/7/6, "Cruiser Rota Etc. – SECRET," undated.

[53] See, for example, GUAS, GD319/12/7/6, Gowan (Palmer) to Brown (Scott), "Pool Abandoned," 18 September 1926.

Price-fixing from 1929

With secrecy somewhat better assured, the WSBC concentrated on developing the terms and conditions under which a rota was to operate. Some rotas, most notably for destroyers and cruisers (given the larger number ordered), lasted for four or even five years, usually far in advance of anything the Admiralty had announced, such was the committee's faith in the scheme's future operation. Unfortunately, few examples before 1934 have survived. Since the most complete evidence of such a scheme dates from 1935 onwards (and there is little to suggest that the same basic principles had been altered substantially compared with some years beforehand), examining later records can shed light on what rules were imposed upon firms for its duration. In the case of the 1936 destroyer rota, it is worth quoting the first four clauses in full:

> Consequent upon the enlarged prospective Admiralty programmes in the next few years and prospective foreign enquiries, the following revised arrangements in connexion with the Rota are agreed:
>
> 1. The Rota shall be continued until completion of the round that is in progress on 31 March, 1940.
>
> 2. During the four years ending 31 March, 1940, Denny, Thornycroft, White and Yarrow shall all receive orders for an average of two destroyers per annum, either British or First-Class foreign, and they undertake to accept such foreign orders if required to do so by the Rota arrangements.
>
> 3. Vickers-Armstrong, Cammell Laird and John Brown will each forfeit their next turn in the Rota, following their Argentine orders.
>
> 4. The current enquiry will be left for Denny, Thornycroft, White and Yarrow to deal with as they mutually arrange, and any order obtained will not be included in the Rota arrangement. The remaining firms will refrain from any further competition on this enquiry.[54]

Although from a time nearer rearmament when orders were more likely, the above document is one of a very small number of surviving records that show how the WSBC allocated contracts among the member firms. As this agreement was a revision of a previous document, it implies strongly that very similar, perhaps even identical, arrangements had preceded it, as the four-year period discussed appears to be simply another "round" of a rota already in

[54] GUAS, GD319/12/7/6, "Destroyer Rota Arrangements – SECRET," 9 October 1936.

operation. Secondly, this rota – and other surviving examples – made clear that some firms were to receive more classes of certain types of ship than others, relative to their size, capacity and areas of expertise. Perhaps most importantly, the clauses above suggest the complete and unchallenged authority of the committee. Apart from a provision in certain circumstances (usually if one yard was unusually busy) for firms to swap contracts, the rota was final and non-negotiable.

How then, could firms be so sure of future orders? Initially, there was the Admiralty's own weak position *vis-à-vis* the Treasury, meaning that there was relatively little inside information of any use to industry: The Admiralty faced similarly uncertain conditions with regards to future budgets and orders. By 1936, however, with at least some form of rearmament on the horizon, the WSBC seemed to be able to plan better. The relationship between senior Admiralty and industry figures allowed for a notable handful within the WSBC to gain information, if not influence, over planned orders. John Barr and Captain Crease of the SEF and Sir Charles Craven of Vickers acted as the go-betweens for information on Admiralty procurement and willingly disseminated what they knew to the group.

In a document pertaining to tenders for capital ships, Barr reported to all tendering firms that:

> It is understood that it is not the desire of the Admiralty that they [Harland & Wolff] should build capital ships ... According to the latest confidential information, the programme of capital ships, aircraft carriers, cruisers and destroyers to be placed between now and March 1940 is as follows ...[55]

The list then went on to detail, by year, how many ships and of what class and type the Admiralty would order. While the name of the source was not revealed, Barr knew, for example, that among other contracts there would be five 8000-ton cruisers and two 5000-ton cruisers in 1937/1938, and there would be nine *Tribal*-class and nine "I-class" destroyers in 1939. The document also had an "assumption" that the Royal Dockyards would not build capital ships or aircraft carriers but would build some cruisers, thus allowing an "attempt to draw up an allocation of work."[56] To have this level of detail four years into the future – even accounting for the more regular orders placed after 1935 – is remarkable and shows a well-developed system of collecting intelligence from the Admiralty and using it to share contracts among member firms. Later, in the war itself, the Admiralty belatedly clamped down on profiteering and caused the WSBC

[55] GUAS, GD319/12/7/6, "Points for Discussion at a Meeting of Firms Capable of Building Capital Ships," 18 September 1936.

[56] *Ibid.*

Table 4.2:
An example of a price agreement between firms for *Warspite* machinery tender

Firm	Estimate	Price to be quoted
Parsons*	£405,000	£448,700
Brown	£387,600	£459,000
Scott	£406,300	£472,000
Cammell Laird	£460,800	£480,000
Wallsend*	£402,000	£467,000
Fairfield	£388,500	£463,000
Vickers	£408,000	£471,000
Hawthorn Leslie	£405,200	£465,000
Mean of estimates	***£407,920***	

*Engine manufacturer only
Source: *Machinery for HMS Warspite (Secret)* GUAS: GD319 12/7/6 (the rounded numbers for the "losing" bids were instructed to "add hundreds" to the final estimate, presumably to give the quotes the appearance of authenticity).

to desperately attempt to cover its own tracks. In other words, after 1930 this suggests that this was almost entirely a one-way relationship.

However, even in 1930 some future information about orders and tenders was available to the WSBC, even if it foresaw only a few months into the future. Nevertheless, this still allowed prices to be easily and effectively fixed. Once a notice of a tender was given to firms, estimates were prepared in an entirely standard way; there was little or no collusion in the first instance, and firms submitted to Barr what were ostensibly honest prices for fulfilling the contract. At a subsequent meeting in London, however, costs were openly shared with all interested parties. The list of estimated prices was then averaged across all firms intending to bid, before a final price to be quoted was agreed upon. A particularly revealing early example is the refit of HMS *Warspite* (see Table 4.2, above). Several points must be highlighted. The first is that the firm providing the lowest estimate, John Brown, was instructed to "lose" the tender by entering a bid higher than Parsons, the eventual winner of the contract. The second is that despite having arranged which firm was to receive the contract, the others did not simply withdraw from the process, but instead submitted higher quotations to give the deliberately false impression to the Admiralty of a genuine bid (indeed, in the handwritten notes at the side of the chart the term "add hundreds" appears next to the losing bids, presumably because the rounded nature of the figures might have aroused suspicion). The third point is that exactly ten percent was added onto the mean of estimates and rounded to the nearest hundred (the exact figure is £448,712), which was the amount quoted for the winning bid.[57]

In adding ten percent to the average, and not simply to the estimate provided by the firm organised to win the contract, the system encouraged other yards to keep their prices competitive. Had this been, for example, Cammell Laird's turn in the rota, a loss would have been incurred on the contract (as the "winning" bid was £448,700, and Cammell Laird's original estimate was £460,800). At the other end of the scale, firms were provided with a disincentive to insert an artificially low price, since it would not aid in winning a contract "out of turn" and driving the average price down would only harm their own margins when it was their turn to "win" in the future. Conversely, had it been John Brown's "turn" to be allocated this contract, the winning price would have represented a mark-up of £61,100, or a profit margin of over fifteen percent on its own estimate. This did not mean necessarily that John Brown was a less expensive or more efficient naval manufacturer than Cammell Laird, as costs could vary substantially depending on the other work the yard had and the materials it held in reserve at the time. Nevertheless, by operating in this way the scheme actively rewarded efficiency, as a firm could not assume it would make a profit irrespective of the initial cost estimate.

Some deductions were made from the final totals. On top of the one percent to cover the tendering expenses, which was split among the other members, money was also deducted for the NSS scheme and the Shipbuilding Conference.[58] Once again, these had an element of secrecy to them. These costs were officially to be hidden from the tender by being distributed throughout the ship;[59] on a complete order, certain amounts would be allocated to the hull, the armour and the engines, so as not to lump several thousand pounds into a single component and possibly draw attention to its inclusion. The NSS, like the other builders, took one percent of the final cost, while a much smaller figure – often as small as a few hundred pounds – was allocated to the Shipbuilding Conference.[60] While these costs inevitably ate into profits, they were a small price to pay given the general industrial uncertainty of the period. In the case of *Warspite,* the total deductions were calculated at £10,500, or 2.3 percent of the final cost.[61]

It should be borne in mind that profits of around ten percent represented very healthy returns before rearmament – average profits in the merchant and passenger sector remained perilously low well into the 1930s. For example, in late

---

57 GUAS, GD319/12/7/6, "Machinery for HMS *Warspite*."

58 GUAS, GD319/12/7/6, Craven to Brown, 20 April 1931.

59 GUAS, GD319/12/7/6, Craven to Brown, 21 April 1931. This letter split £25,000 of the tenders cost as £13,000 to the hull, £1,000 to electrical equipment and £13,000 to machinery.

60 GUAS, GD319/12/7/6, "Shipbuilder Codenames as of 1932 *SECRET: Minutes of Private Arrangement made by letter concerning 'Minespools 32'*."

61 GUAS, GD3192/7/6, "Machinery for HMS Warspite' *(Secret)*."

1931, the price quoted for a standard 7500-dwt cargo steamer had fallen by more than £11,000 to £37,500 compared with two years earlier.[62] Even this reduction failed to stimulate demand; work was still scare, and unemployment still high. Even when contracts were secured, market conditions were unfavourable – Fairfield had problems with customers defaulting on merchant and passenger ships contracts as late as 1935.[63]

Moreover, fixed-price contracts like these should normally have carried a high degree of uncertainty for the manufacturer: any unforeseen workplace stoppages, such as strikes, or unexpected peaks in the cost of raw materials could have eradicated any projected gains, while attempting to build these into the bid for a tender and increasing the price ran the risk losing out on the contract altogether.[64] Naval yards should in these circumstances have been competing furiously for a relatively small number of Admiralty contracts to keep money coming in. The committee's work almost entirely eradicated such uncertainty.

Thus, membership in the WSBC paid handsome dividends – financial and otherwise – even in the harsh climate of disarmament. Moreover, the very existence of a committee which operated to rig the prices paid by the Admiralty, its main customer, is doubly shocking insofar as the public belief and common contemporary criticism of private arms manufacturers was that they had an uncomfortably *close* relationship with the various governments, MPs and Admiralty figures of the day.[65] In this case, the public were right about the existence of unethical practices in armaments manufacture, but wrong about the nature of it.

Considering the above information, there are three plausible hypotheses for how much the Admiralty knew about the existence of the WSBC. The first is that it did not know anything at any point about the WSBC. This is highly unlikely. In 1925, there was simply too much overlap between industry and Admiralty interests, manifested in the controversial Chatfield subsidy, to believe that a parallel operation involving some of the same firms and actors was formed with absolutely no Admiralty knowledge. This leaves either the second option, namely that the Admiralty had known about the price "ring" all along, or finally that somewhere between 1926 and 1936 the relationship changed and industry decided to "go it alone." The third scenario is the most likely. For while there were obviously good reasons to share work in the climate of 1926, there were far fewer reasons for allowing an effective cartel to be extended to all types of ships once recovery was under way. It must be supposed that had the Admiralty known about the *extent* of cooperation, it would have acted to break the ring up at some point before or during rearmament. However, the WSBC firms were so

62 Slaven, *British Shipbuilding,* 91.

63 Peebles, *Warshipbuilding,* 142.

64 W. Ashworth, *Contracts and Finance* (London, 1953), 107–108.

65 W.H. Williams, *Who's Who in Arms* (London, 1935), 14–15.

fearful after the outbreak of war of the Admiralty discovering the collusion that they attempted to twist the figures and present their costs in the most favourable light.[66] This is hardly behaviour one might expect from a group fully in cahoots with its main customer.

The National Shipbuilders' Security Scheme, 1929–1931

In sum, the WSBC was a rather dramatic response from industry to the crisis in naval manufacture following earlier unsuccessful attempts to secure a stable income stream. It was not the only industry-led development of the 1920s which had an impact specifically on the private manufacture of naval armaments, however, nor was it the only initiative that had a transformative effect on the relationship between state and industry. In 1929, a separate but related development, Sir James Lithgow's National Shipbuilders' Security (NSS) scheme, began to gather momentum.

In its simplest terms, the NSS was a project by which firms could voluntarily offer themselves up for sale to the backers of the scheme so their yards would be "sterilised" and not made available for future shipbuilding use. The term "sterilise," which the NSS itself used, was an accurate representation of what the scheme sought to do: when a yard was purchased, it was stripped out and demolished. Moreover, a legal clause prohibited a new owner from using the site to build ships for an extended period, typically several decades.[67] The rationale behind this was straightforward. The fear was that unless such measures were taken, the yard might be reopened at the first sign of a market upturn, thus making it impossible to fully solve the underlying problem of overcapacity. Although the NSS worked mainly in the merchant and passenger sectors, it is of primary interest to this story through the impact it had after the Manchurian incident on the supply of naval arms and the CID's changing defence hypotheses. Indeed, the central involvement of Sir James Lithgow – a key figure in war planning in the late 1930s – in the NSS is crucial. Moreover, Lithgow's business acquisitions of naval armaments firms in the later 1930s, including Beardmore and Fairfield, take on new significance when coupled with his work for the CID, and leads to the conclusion that the NSS deserves much more attention in the history of British naval armaments and shipbuilding than it has hitherto been accorded.[68]

---

66 GUAS, GD319/12/7/6, Crease to all Warship Group Firms, 13 October 1941.

67 NMM, "Prospectus, National Shipbuilding Security Ltd.," 1930.

68 The NSS has often been consigned to footnotes in the histories of British industry, most frequently in the context of rationalisation in shipbuilding following the downturn and depression which characterised most of the 1920s and early 1930s. Professor Tony Slaven has been one of the very few (Professor Hugh Murphy is another) to afford the NSS the attention and significance it deserves as an organisation critical in reorganising shipbuilding, and he has been one of a tiny number to utilise the papers of the NSS. Much

Sir James Lithgow co-owned with his younger brother Henry the Port Glasgow-based shipbuilding firm Lithgows Ltd. (his father, William Todd Lithgow, was a co-partner in the yard with Joseph Russell, and the yard was known as Russell & Co. until renamed in 1918).[69] James Lithgow was not just a well-known figure on Clydeside; he had also served with some distinction in the Great War, seeing action at the front and rising to the rank of Colonel. Later in the war he was called back to London to serve as the Director of Merchant Shipping. He subsequently was knighted, served as the British employer's representative on the International Labour Organisation (ILO) in Geneva and for three years was President of the Federation of British Industry. His upbringing on the Clyde had brought him into contact with the engineering magnate and PSOC architect Lord Weir before the Great War, and thereafter Lithgow considered Weir a "mentor."[70] He had also developed a close bond with another Glaswegian, Sir Andrew Rae Duncan, who had held senior appointments in the Clyde Shipbuilders Association, the SEF, Imperial Chemical Industries (ICI) and the Bank of England. Indeed, Lithgow's biographer later noted that Lithgow made "several important decisions in his life based partly on the advice of Lord Weir and Sir Andrew Duncan."[71] Lithgow was nevertheless a respected, influential and well-connected figure in his own right, and with Weir and Duncan formed a formidable trio of industrial magnates.

Lithgow and Duncan had long harboured a desire to boost the shipbuilding industry. As such, the NSS should be viewed against the backdrop of generally unsuccessful attempts to encourage production in merchant and passenger construction throughout the 1920s, but took on a new direction under Lithgow (namely, to restrict capacity instead of just stimulating demand). State intervention in shipbuilding had been deeply unpopular among firms for many years before the 1920s, and the idea of nationalising the industry – mooted as late as November 1918 – set off vigorous reactions from the SEF whenever it was brought up.[72]

---

has been written up for the first time in his book, *British Shipbuilding,* published in the winter of 2013 as this book was in its final stages. Unfortunately, the archives of the NSS were lost some decades ago, apparently during the transfer of the National Maritime Museum (NMM) archives from the former Woolwich Arsenal to the Royal Naval College, Greenwich. What survives today is largely based on the extensive notes he took of the collection in the 1970s, cited by Professor Slaven in the format "National Shipbuilders Security, Deeds of Covenant, 27 February 1930, NMM," or the correspondence with the NSS surviving in the records the individual firms themselves. Where not otherwise cited or given as from company records, what follows is based on these notes and is used with Professor Slaven's permission.

69 H. Murphy, "Déjà Vu all over again: The Reluctant Rise and Protracted Demise of Scott Lithgow" (PhD thesis, University of Westminster, 2001), 33.

70 J.M. Reid, *James Lithgow: Master of Work* (London, 1964), 90.

71 *Ibid.*, 93.

72 C.E. Fayle, *The War and the Shipping Industry* (Oxford, 1937), 337.

Of course, before 1922, the fortunes of yards and the forecasts of future prosperity rendered any such move highly unlikely, yet even the poor market conditions that persisted thereafter did little to bring many in the industry over to the idea. Thus, in spite of the wide discrepancy between demand and the number of available berths, in part as a result of wartime expansion and the limited orders for ships, the President of the SEF, Sir William Currie, declared in 1929 that the "Shipbuilding Industry has always been opposed to State aid or interference of any kind."[73] On this point, the SEF was at least in luck: government policy had shifted markedly since the conclusion of the war, and it had no intention of intervening in industry either.[74]

Continued opposition to interference should not be confused with the notion that shipbuilders would refuse to accept any help from the state. As Slaven notes, while "intervention was anathema to [the industry], friendly support without strings attached was another matter."[75] Indeed, comparisons were frequently drawn in the 1920s by shipbuilders with Germany, Japan, the United States, Italy and France – these governments all either subsidised ship construction or undertook some form of fleet modernisation. In this regard, British shipbuilders had grown increasingly keen on such "friendly support" after the boom passed in 1921.[76] The closest British equivalent was the Trade Facilities Act (TFA), also of 1921, which provided money to guarantee loans for any capital project, including shipbuilding. As the system guaranteed the loans taken out by private buyers to cover the cost of the vessel they purchased, it was therefore inapplicable to naval shipbuilding (for the obvious reason that the buyer was the state itself).

More importantly, the TFA was also poorly conceived and not particularly successful in the context of shipbuilding. Since overcapacity in berths and an oversupply of merchant vessels were the main problems, the British scheme – coupled with those abroad – only exacerbated the sector's existing issues. Instead of helping to rationalise the industry, the TFA merely tended to prop up the already excessive number of otherwise uncompetitive yards and to keep prices (but not costs) low, which only further hindered the small and dwindling number of naval firms from filling the holes in their order books with merchant or passenger ships.[77] Thus, although popular with struggling firms, the political will to keep the act waned. The Chancellor of the Exchequer, Winston Churchill, believed by 1925 that the act had outlived its usefulness (though he did not have the same perceptiveness about the Gold Standard), while Prime Minister Baldwin believed that the financial situation necessitated scrapping the scheme.

[73] *Glasgow Herald Trade Review*, 1929.

[74] Slaven, *British Shipbuilding*, 85.

[75] *Ibid.*

[76] By far the best overview of this topic is to be found in L. Johnman and H. Murphy "Subsidy and Treasury: The Trade Facilities Act and the UK Shipbuilding Industry in the 1920s," *Contemporary British History*, 22/1 (2008), 89–110.

[77] *Ibid.*, 102.

He also suggested that giving "special facilities to a competitive industry" only encouraged other nations to respond in kind.[78] Thus, the act was terminated in March 1927.[79]

The abolition of the TFA prompted Lithgow, Duncan and others to seek alternatives, and it was through this that the NSS was born. The NSS did not therefore represent the first attempt to assist struggling yards, but it was a notably different direction in that industry, not government, took the lead. This new scheme was not brought into existence immediately; for the first year after the TFA was abolished, market conditions did not deteriorate markedly, although new orders remained static. Similarly, although the unemployment rate in shipbuilding actually fell markedly, it continued to remain far higher than the national average and never returned to 1921 levels.[80] By 1929, however, there was mounting evidence that another large-scale deterioration was imminent; work in progress had fallen by over forty percent in a year, and this was exacerbated substantially with the global financial panic that started that autumn on Wall Street.[81] Not only were the prospects for merchant and passenger work in the near future bleak, but the renewal of arms limitations at the 1930 London Naval Treaty meant that expanding warship production remained out of the question. As state assistance, friendly or otherwise, was not forthcoming, the NSS scheme for rationalisation – with rather bolder aims than the TFA – was hatched.

The NSS grew directly from the discussions being held in the Shipbuilding Conference over reviving the industry's fortunes. By the end of 1928, the Conference believed that the best course was to instigate a scheme for spreading the small amount of work available around fewer yards, since "uneconomic competition" through artificially low prices constituted more of a threat to the industry than did the collapse of firms. This scheme was agreed in principle by March 1929 and developed from that point until its official launch in 1930.[82] The remaining issue was how to pay for it, since it would inevitably involve purchasing and liquidating a great deal of shipbuilding capacity. To this end, Duncan, former president of the SEF and by 1929 a director of the Bank of England, wrote to the Bank's Governor Montagu Norman on behalf of the conference with a request to provide banking facilities and to underwrite the NSS.[83]

Duncan did not have to work especially hard to convince Norman of the merits of the scheme, as the Governor had been a vocal supporter of rationalisation schemes for some time. Thus, with Duncan's help it took only a few weeks

---

[78] TNA, T160/550/8, "Trade Facilities Acts: Deputation from Employers in Shipbuilding Industry," 15 February 1927.

[79] M. Lindberg and D. Todd, *Anglo-American Shipbuilding in World War II: A Geographical Perspective* (Westport, CT, 2004), 75.

[80] Johnman and Murphy, *British Shipbuilding and the State,* table 3.

[81] Slaven, *British Shipbuilding*, 88.

[82] *Ibid.*, 95–96.

[83] BOE, SMT/2/280, file 142, Securities Management Trust, Duncan to Norman.

for Norman and other senior Bank figures to meet a small Conference delegation led by Lithgow and John Barr (now controller of the Shipbuilding Conference and the WSBC). There was no government involvement. Here, Lithgow asked Norman for assistance in providing capital to support the new scheme, with a figure of around £2,000,000 mooted. Norman agreed to assist but stopped short, for the time being, of an agreement to provide the money from Bank reserves.[84] He also stressed that any support would be on the condition that no vast conglomerates were formed in the process which might reduce "healthy" competition. He was unequivocally assured that uneconomic competition was the gravest danger yards faced and that the new scheme was not intended to create larger, unwieldy companies.[85] However, given John Barr's central role in organising the WSBC (which in one way or another was running a kind of conglomerate that eliminated "healthy" competition) it could be suggested that in the case of warship yards, this agreement was worthless.

Another Bank of England director, J. Frater Taylor, advised Norman that there was no "reason why the [industry] could not be put on its feet" if there was a "live force behind the project" and suggested that Norman keep "pounding at Lithgow" to ensure he stayed involved and focused on the project. It was partly because of Norman's encouragement that Lithgow soon emerged as the "live force" he was seeking to rationalise the industry, and the two discussed their ideas frequently in the second half of 1929. Lithgow's connection to Norman and Duncan, coupled with his active role in persuading firms around Britain to back the scheme was crucial,[86] and it was thus no surprise when, at the first meeting of the NSS in the weeks following its constitution in February 1930, Lithgow was unanimously elected Chairman.[87]

The NSS, it was agreed, existed for "the purpose of purchasing redundant and obsolete shipyards, the dismantling and disposal of their contents and the resale of sites under restriction against further use for shipbuilding."[88] By restricting the resale and reuse of sites for shipbuilding for a period of many (typically forty) years, the NSS attempted to ensure that yards would not re-enter a fragile market at the first sign of recovery. Unconventionally, it was up to the owners of the yards to declare their firms redundant and to liquidate themselves voluntarily by selling their assets for a "fair" market price, which was determined by John Barr, who was appointed by the NSS as valuer.[89] Barr's close involvement with the NSS, the Shipbuilding Conference and the WSBC illustrates the overlap between the membership of three ostensibly separate organisations, although while the NSS

84 BOE, SMT/2/280, file 142, "Meeting with Norman and J. Frater Taylor," 29 October 1929.

85 BOE, SMT2/280, file 142, "Meeting with Shipbuilding Conference," 24 April 1929.

86 Slaven, *British Shipbuilding*, 97.

87 NMM, NSS, "Meeting of Proposed Subscribers," 28 March 1930.

88 Slaven, *British Shipbuilding*, 97; NMM, NSS, "Deeds of Covenant, 1930."

89 Murphy, "Déjà Vu all over again," 48.

had similar motives to the WSBC, in that both were concerned with excessive competition, each responded to the challenge in entirely different ways.[90]

The NSS began life on 27 February 1930 backed by a clear majority – forty-six – of the largest shipbuilding firms, including almost all the WSBC members. Its board of directors included Sir Charles Craven of Vickers, Sir Alexander Kennedy of Fairfield, Sir Frederick Rebbeck of H&W and a number of smaller merchant shipbuilders (including Sir Amos Ayre, then director of the Burntisland Shipbuilding Company, but later the chairman of the Shipbuilding Conference and Director of Merchant Shipbuilding during the Second World War).[91] Perhaps sensing the possibility of allegations of conflicts of interest and the unpopularity of a scheme which closed down yards in areas already experiencing high unemployment, the board immediately applied (successfully) for an exemption to remove the names of the directors from the new company's correspondence.[92]

The NSS had to proceed without having finalised financial arrangements with the Bank of England, although discussions were still ongoing. Instead, the scheme only had the comparatively minor sum of £100,000, raised through an issue of £1 shares – far short of the £2,000,000 Lithgow believed it would need to fulfil its goals. This lack of capital meant that when the NSS made its major purchase, paying £209,000 for the shipyard of the giant company (and WSBC member) Beardmore, it had to go to the Bank of England to borrow the money (secured against the assets of the backer firms) to cover the cost.[93]

The purchase of Beardmore had wide significance. It had once been the fourth largest British shipbuilding firm in terms of market capitalisation,[94] and with thirteen berths with a theoretical construction capacity of over 100,000 gross tons, it was also (apart from Palmers purchase by the NSS in 1934) the largest yard to be sold to the NSS.[95] It was closely connected to the Bank of England, making it a rather unusual case which requires some explanation. Less than three years before the sale, the Bank had provided the Beardmore company (which included the steel forge in Parkhead in the east end of Glasgow) with three-quarters of a million

90 The WSBC had a small membership of firms that had unique capabilities and, most importantly, were the only ones used by the Admiralty for naval orders. Thus, the barriers to entry were so high that an effective rota could be devised for naval orders. Given that the rest of industry was not dealing with the state but instead with private buyers in Britain and abroad for more general classes of ship, such a rota was clearly not possible, and competition had to be removed in other ways. It did not preclude WSBC members being part of the NSS, as their business interests overlapped with passenger and merchant work.

91 L. Day and I. McNeil (eds.), *Biographical Dictionary of the History of Technology* (London, 2013), 55.

92 NMM, NSS, "Minutes of Meeting," 15 May 1930.

93 Johnman and Murphy, "Welding," 96.

94 Slaven, *British Shipbuilding,* table 3.5.

95 *Ibid.*, table 4.9.

pounds to see it through its troubles in return for an equity stake in the company. The scheme devised by Norman and Duncan was for the money be used to convert existing loans to other banks – which had little chance of being repaid otherwise – into cumulative preference stock, effectively giving the Bank complete voting control over the firm, which in turn was used to oust the incumbent chairman Lord Invernairn. Interestingly, in 1928 Norman had offered Lithgow (and considered offering Weir) the chairmanship of the firm but, after Lithgow declined the offer to focus on his own yard, Norman consulted Duncan and instead settled on the former director of Coventry Ordnance Works, H.A. Reincke.[96]

By 1930, however, Beardmore's financial situation was dire, and the Bank's money was inadequate to stop the blood-letting. The scrap value of the shipyard in the very depressed market was barely £30,000.[97] Thus, while the £209,000 the NSS paid to liquidate the shipyard went only a quarter of the way to recouping the investment – and gave up by far the larger part of the company in doing so – it was still a good price given that there were no other interested buyers, and it allowed the Bank to hold on to the marginally healthier steel forge in the process. Indeed, Montagu Norman's secretary noted that "nothing like these terms would have been obtained had it not been for the interest of [Norman] in the [NSS]; this is one way in which Lithgow has shown his gratitude."[98] In other words, Norman's financial support for Lithgow's scheme – even if it meant loaning the NSS money to buy the yard from the Bank – allowed for a smooth sale at a price which satisfied the owners, satisfied the Bank, and removed substantial capacity from the market, thus satisfying the industry.

While it could be suggested that a "cheaper" option for the firms which backed the NSS would have been to allow Beardmore to fail, it should be borne in mind that such a course of action would have had adverse consequences. First, an orderly sale was better for the industry's public relations and investor confidence than an enforced collapse of one of its industrial giants. The Bank of England well understood this.[99] More importantly, however, buying Beardmore allowed the NSS to insert the key clause for the restriction of the firm's shipbuilding capacity for a period of forty years, whereas liquidation alone offered no guarantees that the huge Dalmuir yard would not re-enter the market when conditions even marginally improved.

Of course, liquidating major industrial concerns, even if deemed beneficial for the industry, was not without side effects. For one, closing yards could be very unpopular in already depressed communities.[100] To allow the work of the

96 Hume and Moss, *Beardmore,* 212.

97 Johnman and Murphy, "Welding," 96–97.

98 BOE, SMT/2/280, Memorandum, 30 May 1930.

99 BOE, SMT/2/280, "Notes and Memoranda of November 1929."

100 The "Jarrow March," where hundreds of unemployed workers marched from the North East to London to protest following the closure of Palmers is perhaps the best known.

NSS to be conducted as efficiently as possible and to avoid allegations of conflicts of interest, Lithgow, in addition to removing the public listing of directors, repeatedly demanded absolute secrecy from his board. Members were instructed to "refrain from talking" to the press and told that all external communications, even at AGMs, were "at the discretion of Sir James Lithgow."[101] It was only in 1932, in the face of mounting criticism of the NSS's work, that Lithgow changed tack and declared that some "advertising" of the scheme's merits might be beneficial. By 1933, Lithgow went further still, indicating that while it was best "not to seek chances to enter controversy," board members should "if the occasion arose ... not miss the chance to explain the aims of the company."[102]

In the earliest days, the NSS simply needed funding. To this end, a hugely significant milestone was reached on 21 January 1931, two months after the Beardmore deal was concluded, when £1 million in loan guarantees was offered to the NSS through a subsidiary of the Bank known as the Bankers Industrial Development Company (BIDC). The Bank had set this project up the previous year to finance schemes to reorganise industry. The loan was to be made as First Mortgage Debenture Stock, redeemable over thirty years, with an annual interest rate of five percent.[103] While the BIDC was not aimed solely at shipbuilding, the £1m outlay represented a sixth of its available funds, making the NSS one of the largest beneficiaries.[104] To pay the interest on the loan, or if necessary to raise extra capital for the purchase of more yards, the forty-six firms backing the NSS agreed to pay one percent on the sale price of all contracts undertaken, back-dated to the end of 1930. With this support, the NSS could purchase yards at a tremendous pace; within barely eighteen months of the Beardmore sale, the total number of berths liquidated reached eighty-two with a total capacity of 491,000 tons.[105]

Before jumping too far ahead in the narrative, it is important to highlight an issue referenced above: how compatible the Admiralty's grave concerns over maintaining naval shipbuilding capacity and capabilities were with these broader industry rationalisation measures. As already noted, the technological differences between a warship and a cargo vessel meant that the problem of idle plant for equipment such as armour and guns was very different from that of an idle slipway in a merchant shipbuilding yard. Thus, the ideal solution to the former problem for the Admiralty was to find a way to sustain excess capacity should it be required at some future date, while the ideal solution for the industry was to find a way to eradicate excess capacity, preferably for a very long time. These competing motivations could not easily be reconciled, and the NSS therefore from its inception conflicted with the Admiralty's wishes to see capacity

---

101 NMM, NSS, Minutes of meetings, 15 May 1930 and 28 January 1932.

102 NMM, NSS, Minutes of meeting, 12 October 1933.

103 Slaven, *British Shipbuilding,* 98.

104 R.S. Sayers, *The Bank of England, 1891–1944* (Cambridge, 1976), 229.

105 NMM, NSS, Minutes, 28 January 1932.

retained. Clearly, the firms represented in the WSBC existed as businesses first and foremost; if it made more sense to liquidate capacity through the NSS, as it did for Beardmore, then concerns about the implications for a future war effort did not feature in the decision making, at least in 1930.[106] As the next chapter will show, however, Manchuria changed this dynamic.

Conclusions: The WSBC and NSS in the Context of Shipbuilding before Manchuria

After Washington, the Admiralty, along with every shipbuilding company, naval or otherwise, had to adjust to new, unfamiliar and uncomfortable post-war conditions. Of these groups, the private naval arms industry was least able to adapt. Following several unsuccessful attempts by both sets of actors to lobby for subsidies or other support, the WSBC was an unintended consequence of government's lack of foresight about private industry. Its evolution was a direct result of the Admiralty's belief both it and private industry shared the same interests. The reasons for the WSBC going it alone can be boiled down to powerful motivations for self-preservation first and profits second. There never was a complete breakdown in Admiralty-business relations. The WSBC's members outwardly maintained close working ties, frequently sharing staff and information, while at the same time conspiring to form a ring that fixed and inflated prices.

The WBSC came into existence and carried out its operations because the entire industry sat outside the corridors of political power. It was not an effective lobby group in the 1920s and could not alter political currents. The WSBC thus came to exist through *lack* of influence as a *defensive* group that was formed for self-preservation rather than patriotism or loyalty to its main customer. Here, too, it is worth underscoring that the aims of the government and those of the military differed substantially after the war, and the two cannot be thought of as part of a single, monolithic state with a unified vision of defence. Indeed, in many ways, between the Washington Treaty and Manchuria, the Admiralty was as much an outsider as industry.

The Manchurian crisis completely shifted this landscape. By the end of 1933, some industrialists – although not WSBC members – even sat on subcommittees of the Committee of Imperial Defence and advised on the manufacturing and defensive "deficiencies" that faced Britain. Thus, we must now turn to explain how the shift in British (and global) politics between the summer of 1931 and the spring of 1932, starting with the formation of a National Government, changed the debate on defence, and the effect this had on private industry.

---

106 During the discussions for the sale of Beardmore, the question of Britain's "need" for its naval plant was not a factor. See NMM, NSS, Minutes, 15 May 1930.

Part Three

The Formation of a National Government, the Far East, and the PSOC Approach to Industry, 1931–1934

Chapter Five

From "Outsiders" to "Insiders": Industry and the Rehabilitation of the Supply Planning Framework

The three politically tumultuous years that followed the formation of the National Government in 1931 fundamentally transformed the nature of defence planning and the relationship between government and industry. Between 1922 and 1931, the debate between Admiralty and Treasury had been comprehensively won by the latter's view on spending limitations. The Committee of Imperial Defence (CID) and its subcommittees had not been called upon to work on pressing matters of national security, but spurred by the changes in the political situation after the Japanese invasion of Manchuria, it adopted a far more central planning role and contributed a great deal towards understanding British defensive deficiencies by 1934. Thus, it was the CID, comprised of all three fighting services, and not just the Admiralty, which became the vehicle for articulating Britain's defence needs to the Cabinet.

This resulted in industrialists playing a key role in advising, and then shaping, defence policy for the first time. On the other hand, by 1934 the Admiralty also began competing more intensely with the developing Air Force for a share of the defence budget. Thus, while the narrative to this point has stressed the private industry's existence outside of the state planning framework and consequently has focussed on industry's responses to the crises in the 1920s and the Admiralty-Treasury disputes, from 1931 onwards the increasing importance of the CID warrants a more detailed examination of the process that gave new meaning to defence planning and that subsequently led to approaching industrialists for assistance.

Overview: Constraints and Pressures

Between Washington and Manchuria, support for the navy had been occasionally vocal but rarely consistent. Winston Churchill argued in favour of increased naval expenditure almost as often as he argued against it before and between 1918 and 1931, first supporting expansion of the fleet before slashing naval estimates and finally implementing the Ten Year Rule for perpetuity.[1] This sort of behaviour was not unique to him: past, present and future Prime Ministers Stanley Baldwin, David Lloyd George and Ramsay MacDonald had periods on both sides of the divide, at times supporting the centrality of the navy to British defence, at others angrily being described by Admiralty figures as "dangerously imperilling everything for which the Royal Navy stands."[2]

This shift in the rhetoric from political heavyweights highlights the severe constraints on domestic, foreign and economic policy under which they were operating. In short, it was acceptable to commit to the principle that the Royal Navy was central to the British way of life so long as it did not mean spending significant sums on it. In MacDonald's case, his Labour party was a mixture of different factions and organisational interests,[3] marked by strong anti-war tendencies, and as such harboured deep divisions throughout the disarmament period, divisions which at all times required careful handling to stop the party from fracturing.[4] Similarly, Baldwin's party had split in 1929 along lines of financial priorities, one of which was how best to divide the armed forces budget.[5] The difference between these years and those after 1931 was that while the domestic political and economic pressures remained (and, in the latter case, intensified after 1929), the geo-political landscape shifted substantially and increased the importance of future defence planning. This was what underpinned the change in the relationship between industrialists, government, and private armaments manufacturers.

The relatively rapid shift from no discernible external military threats in 1929 to a situation where the Japanese, Germans and Italians could all be considered potential enemies by 1935, makes it apparent that this is a period in need of disentangling to understand the changes in defence planning and its impact upon industry. It was only after Manchuria that the CID – and particularly its PSOC – emerged from the shadows to devise methods to tackle deficiencies while the world around it constantly shifted. Within these organisations some industrialists grew into partners with the state in planning war, while others remained in defensive rings to secure the future of their firms. Put simply, it is this period where the "pivot" between total exclusion and limited inclusion of industry occurred.

This section thus considers industry and politics between the fall of the Labour government in mid-1931 and the first report of the CID's "Advisory Panel of Industrialists" in 1934. It considers primarily the formation of a "government of national interest;" Manchuria and the first CID "defence deficiency" programme

1 Bell, *Churchill and Seapower,* 133–134.

2 Neidpath, *Singapore Naval Base,* 64–65.

3 K. Morgan, *Ramsay MacDonald* (London, 2006), 45.

4 Vickers, *The Labour Party and the World,* 80–84.

5 P. Williamson "'Safety First:' Baldwin, the Conservative Party, and the 1929 General Election," *Historical Journal*, 25/2 (1982), 392.

which followed it; and the subsequent approach to civilian industrialists. Four arguments are put forward for the years 1931–1934. Firstly, planning and hypothesising *were* valuable to the later rearmament drive, and this need not be linked just to the policies or programmes that came to fruition. Significantly, whether plans were implemented or not, it was in these years that the initial lines of communication between planners and industry opened and began to function. Indeed, it was in 1932 that deficiency planning started, not – as some historians have suggested[6] – late 1933 once Nazi Germany had withdrawn from the Geneva disarmament conference. Secondly, certain industrialists and civil servants demonstrably benefitted from each other through these channels, and while this has an element of a conspiracy about it, it was in fact much more complex. Thirdly, this period is unique in several ways: the specific nature of the threat in 1931 and 1932, the costs involved, and the time needed to remedy deficiencies meant it was at the outset a *naval* question. Indeed, one could reasonably suggest that had Germany been identified by Britain as the ultimate threat in 1931 instead of concentrating first on Japan, the composition and functioning of the CID's partnership with the small industrial elite might have been entirely different. Finally, any cooperation between industrialists and the CID was possible only because of the structure of the National Government, itself a product of an economic catastrophe unlike anything previously experienced. The National Government's cautious approach and indecisiveness explain the nature and length of defence planning and why there was so little *material* progress in the years after Manchuria. Indeed, this theme of planning without progress recurs throughout the decade.

The Royal Navy and Spending in 1931

From the Royal Navy's perspective, the key issue in the early 1930s was Japan. The Royal Navy was a young and large fleet at the end of the Great War, but by 1930 it had received comparatively little funding for more than a decade.[7] The Imperial Japanese Navy, the third largest in the world, added another dimension to the problem as Japanese interests in China contended with Britain's own in East Asia more generally. Although Churchill had successfully argued in 1925 that the Far Eastern problem did not resemble the kind of crisis Britain faced with Germany before the Great War, by the end of the decade Japanese agitation for increased influence in China was causing renewed concern at

[6] B.J.C. McKercher, "From Disarmament to Rearmament: British Civil Military Relations and Policy Making, 1933–1934," *Defence Studies*, 1/1 (2001), 21–48; and K. Neilson, "The Defence Requirements Subcommittee, British Strategic Foreign Policy, Neville Chamberlain and the Path to Appeasement," *English Historical Review*, 118/477 (2003), 651–684.

[7] Gordon, *British Sea Power,* 73–74.

the Admiralty.[8] In late 1931, the Japanese invasion of Manchuria escalated the "problem" of the Far East into a full-scale diplomatic crisis. Moreover, the key British outpost – the Singapore naval base – had suffered under the same cuts in funding as Britain's private naval manufacturers and had not fully recovered from the cancellation of works in 1924.[9] It was thus underprepared if Britain wished to intervene directly in China.

Underpinning these problems was the London Naval Conference of 1930. Compared to Washington, London was more of a modest revision to existing terms – partly because Japan and the United States would not concede an inch in their respective demands.[10] The most that could be agreed was for an official distinction between "heavy" and "light" cruisers (eight-inch compared with six-inch guns), while tighter restrictions were placed on the sizes of submarines. Perhaps most notably, the 5:5:3 ratio between Britain, the United States and Japan was replaced by a 10:10:7 ratio for all except capital ships (which remained at 5:5:3) and submarines (5:5:5), effectively allowing Japan more room to construct cruisers and smaller vessels while holding Britain and United States at parity. Like Washington, the London agreement allowed the United States and Japan to undertake some increase in construction, as the tonnage limits for cruisers (the United States) and capital ships (Japan) were higher than current levels. The United States used these new limits to partially offset the effects of the depression, a notably different tactic to the disarmament-obsessed Labour party in Britain.[11] This did not, however, put an end to Japanese ambitions in China, although it had the potential to strengthen Japanese naval power relative to Britain's in the region.

Thus, while Britain's naval position in 1930 was broadly like the post-Washington 1920s, the geopolitical shifts, particularly tension in the Far East, added a new urgency to the Admiralty's demands for fleet modernisation. These were still being made, however, in a period of powerful political and economic constraints surrounding disarmament which had existed since Versailles, and which the Labour government had little desire to alter for both ideological and economic reasons. To understand why this scenario changed in the second half of 1931, one must understand the nature of Labour's collapse and the National Government's rise.

---

8 J.H. Maurer, "'Winston Has Gone Mad:' Churchill, the British Admiralty and the Rise of Japanese Naval Power," *Journal of Strategic Studies*, 35/4 (2012), 793.

9 Neidpath, *Singapore Naval Base,* 120–121.

10 *Ibid.,* 118–119.

11 Gibbs, *Grand Strategy,* 30–31.

Origins, Structure and Economic Background of the National Government, 1931: "Not a Coalition in the Ordinary Sense"

The National Government was born from a crisis caused by economic pressure fracturing an already weak Labour government that had been in power since 1929.[12] At no point since Ramsay MacDonald took over as Prime Minister had the private naval armaments industry any cause to believe in his party's business proposals. While unemployment had actually fallen in the year before the Wall Street Crash,[13] the Labour Party had pinned its pre-election hopes on a continuation of this revival and had only proposed to "stimulate the depressed export market," while pursuing the "drastic reduction in armaments" which was "long overdue."[14] This misunderstood and misrepresented the role of naval arms, and rather crudely bundled it together with merchant shipbuilding, assuming that an upswing in one would amply compensate for the other, which it did not. How the government planned to effectively stimulate exports while pegged to gold was another matter entirely. Little had been learned since Washington.

When severe economic problems split the party – but not before the Gold Standard was finally abolished – and precipitated the fall of the government, what followed was a coalition that gained the largest landslide in parliamentary history: 554 seats were won by candidates of all parties backing the new coalition "National Government," totalling over ninety percent of the votes cast, while dissenting Labour MPs who stood in opposition were decimated. The government, initially a temporary emergency response to the economic crisis, was formed through an offer from MacDonald to his Conservative counterparts to stand together for the greater good of the country. While this delivered electoral success, and despite MacDonald returning to lead the new government, over 470 of these seats were held by Conservatives, who had on their own received more than fifty-five percent of the popular vote.[15] The National Government had an undeniable mandate and, crucially, despite the financial restrictions which necessitated the coalition in the first place, was no longer mainly comprised of MPs who staunchly backed disarmament. Moreover, when the crisis failed to pass, the government decided to remain together for a full parliamentary term (and indeed stood again on the same platform in 1935, winning a second term). This unique political situation cannot be underestimated, as it is highly unlikely that industrial elites would have been discussed or approached at the same

[12] Williamson, "Safety First," 408–409.

[13] Slaven, *British Shipbuilding,* 88–89.

[14] "Labour's Appeal to the Nation" (Labour Party Election Manifesto, 1929), Warwick University Digital Archive [http://contentdm.warwick.ac.uk/cdm/ref/collection/tav/id/1781, accessed 19.11.2013].

[15] H. Berkeley, *The Myth that Will Not Die: The Formation of the National Government* (London, 1978), 116.

time or in the same way had the previous government emerged intact from the economic crisis of 1931.

This event is important for war planning when taken in tandem with the events unfolding in Manchuria, because Manchuria produced a naval threat without removing the restrictions set by the London Treaty or economic constraints. Indeed, the worsening economic situation had already had grave implications for the Admiralty's budget. Naval expenditure had been cut in 1931 to its lowest level since the war, £52.5m. This was £5.7m – or ten percent – lower than the last year of the previous Conservative government, and while it still represented over forty percent of the total sum available for all three services, a cut of this magnitude in peacetime coupled with the overheads reserved for the administration and salary costs of the Admiralty (significantly, more than half of the total sum) meant the money available for new ships was slashed almost to the bone.[16] When the First Lord of the Admiralty, A.V. Alexander, introduced the navy estimates to Parliament, he noted that:

> I have been pressed by the shipbuilding industry on more than one occasion recently that, in view of the depressed state of that industry, we should anticipate naval building of the next few years; but both from the point of view of the financial position and from the fact that such an expedition of naval building would prejudice the negotiations for disarmament, it is impossible for me to accede to such a request.[17]

Thus, 1931 was the second-worst year for naval construction between the wars, but also one when a new naval threat emerged.[18]

Japan, Manchuria and British Naval Defence

In the period between the National Government's inception in late August and election success in October, Japanese forces invaded the Manchuria region of North Eastern China, annexing it and declaring a puppet state.[19] Japan, like the Britain, the United States and Germany, had been badly crippled by the ongoing economic crisis.[20] The downturn in international trade was exacerbated when China, Japan's second-largest market, boycotted Japanese goods. The boycott

16 See Johnman and Murphy, *British Shipbuilding and the State,* table 6.

17 Hansard, HC Deb, vol. 249, c. 1209, 11 March 1931.

18 *Ibid.*, 55–57.

19 Gibbs, *Grand Strategy,* 75.

20 M. Shizume, "The Japanese Economy during the Interwar Period: Instability in the Financial System and the Impact of the World Depression," *Bank of Japan Review*, 2 (2009), 4–5.

was directly linked to the question of Manchuria, where the Chinese government alleged that Japan had encircled the important transport (and economic) artery, the South Manchurian Railway. The crisis began in earnest in September 1931 after Japanese forces made a swift invasion to secure their country's political and economic interests.[21] Over the following months the situation worsened, and Japan used an attack in Shanghai on Japanese nationals by a Chinese mob in January 1932 as a pretext to strengthen naval forces around the city, leading to open conflict until March. While Manchuria was undoubtedly of wider international significance, it was the resultant naval blockade near Britain's interests in Shanghai which was the most serious aspect of the problem.[22]

The National Government now found itself in a difficult position. Public opinion both in Britain and internationally demanded that the League of Nations impose sanctions (although few were in favour direct of military action) against Japan, while British ministers were anxious to avoid any moves which could entangle Britain in a complex problem that might lead to further military involvement in the Far East.[23] The reduced strength and poor state of British defences in Hong Kong and Singapore was a primary factor in the British decisions which followed. As hostilities continued around Shanghai, the Permanent Under Secretary of the Foreign Office, Sir Robert Vansittart, noted in a memorandum that "we are incapable of checking Japan in any way if she really means business and has sized us up, as she certainly has done … Therefore, we must eventually be done for in the Far East."[24] This assessment of British weakness in the face of new threats was highly significant in the CID deliberations that followed.

Vansittart's view was reinforced in a report by the CID Chiefs of Staff (CoS) – a subcommittee comprising the First Sea Lord, the Chief of the Imperial General Staff and the Chief of Air Staff – who believed that even if emergency action was taken the naval bases would fall before a fleet capable of defending them could be dispatched to the area.[25]At a series of meetings in March. the Cabinet agreed that while the situation was alarming, Britain was not militarily, economically or financially prepared for a war in the Far East unless assistance from the United States – a highly remote possibility – was forthcoming.[26] Essentially, Britain was forced into conceding Japan's "legitimate claim" in China because it was unable to make the naval commitment to deter or engage the Japanese fleet.[27]

21 Gibbs, *Grand Strategy,* 74.

22 Neidpath, *Singapore Naval Base,* 123.

23 *Ibid.*, 124.

24 DBFP, II, No. 282, Vansittart, minute, 1 February 1932; and I. Nish (ed.), *Anglo-Japanese Alienation, 1919–1952* (Cambridge, 2010), 39–40.

25 Vansittart, minute, 1 February 1932.

26 Gibbs, *Grand Strategy,* 79.

27 Neidpath, *Singapore Naval Base,* 124.

The complex web of foreign policy that surrounded the Far Eastern question – like the composition and workings of the National Government in Britain – is well known to diplomatic and political historians, and space does not exist for discussing it in full here. The important fact worth noting is the change that Manchuria caused in defence planning. Manchuria highlighted Britain's defensive deficiencies, and the National Government clearly started taking such deficiencies seriously, although it lacked the economic means and public support for large-scale rearmament. The first *Statement Relating to Defence* (or White Paper on defence), which alerted the public and much of the business community to the possibility of rearmament, was not published until March 1935, more than three years after Manchuria.[28] While this was undoubtedly a crucial moment in the history of defence between the wars, this gap is important because what happened between January 1932 and March 1935 at the CID was therefore shrouded in secrecy, and not reported in the press or known by the vast majority of industrialists. Indeed, the term "planning war, pursuing peace," which has been applied to America between the wars, is also applicable to the years between 1932 and 1935 in Britain.[29] The public, through organisations like the National Peace Council and the League of Nations Union, which together had hundreds of thousands of subscribers, remained broadly against rearmament and armed intervention before 1935, which only served to further constrain the government's freedom for political manoeuvre. Thus, the need for effective defence planning – preferably by spending as little money as possible in a way that attracted as little attention as possible – was paramount.

Abandoning the Ten Year Rule and Restarting Industrial Cooperation

The first, and most notable, effect of Manchuria on policy was the beginning of the process which led to the cancellation in March 1932 of the "Ten Year Rule" which had governed strategic thinking since 1919. The assumption that there would not be a major war for at least a decade – renewed annually until Churchill made it self-perpetuating in 1928 – was believed by the CoS to be outdated and in need of revision. The subcommittee produced a report which warned Cabinet that, in the event of conflict in the Far East, British defences were in such a sorry state that they were (with special reference to the Singapore base) "completely inadequate."[30] For instance, the report noted that no vessel larger than a light

[28] Cmd. 4827, *Statement Relating to Defence*, March 1935.

[29] P.A.C. Koistinen, *Planning War, Pursuing Peace: The Political Economy of American Warfare 1920–1939* (Lawrence, KS, 1998).

[30] TNA, CAB23/70, CoS 295, "Imperial Defence Policy," Annual Review by CoS, 23 February 1932.

cruiser was stationed in Singapore, Hong Kong or on the Huangpu River, the three stations most likely to be involved in any action. The report also warned that if these bases were lost, it would end British naval supremacy east of India, leaving "our vast territorial and trade interests in the Far East as well as the coastline and communications with our Dominions and India, open to attack."[31]

Although these bases were nearly six weeks' sailing from Britain – and as such needed to be able to withstand an attack for a longer period without naval support – the report was also critical of the state of defences across the board. The CoS stressed that throughout the Empire naval defences were scarcely capable of accomplishing the most basic tasks of protecting critical trade lines and communications. Moreover, the problem was not only confined to bases but also to the number and quality of fighting vessels. The Ten Year Rule, the report's authors argued, had been an "insurmountable barrier" to taking necessary steps to rectify the problems facing the Royal Navy and wider British Imperial defence, and its continued adoption for the previous thirteen years had lulled the public into a false sense of security regarding the potential for conflict and ignorance as to the condition of the armed forces.[32]

The CoS report was placed in front of the CID and quickly accepted in March 1932. The report brought into focus the "mistakes" that had allowed bases and ships to become obsolete in the decade since 1922. Secondly, Britain relied on accumulated stocks of weaponry, which minimised expenditure and had a degree of economic and political rationality in the 1920s, but also had profound side-effects. Thus, when the Chiefs of Staff highlighted the deficiencies in 1932, they not only had to convince the Treasury to take immediate action but also had to account for the erosion of industry and skills over the previous decade.[33] Japan was therefore not conceptualised as an imminent or major threat to British security even after the incident in Shanghai; the report was as much about the long-term decline of defensive capabilities as it was about war. Manchuria is better thought of as an event which prompted a re-evaluation of the condition of British defences across the board and brought the question of armaments back to the table for the first time since the cruiser crisis seven years earlier. The larger and longer-term impact of Manchuria was to start the ball rolling on the question of deficiencies and to bring much greater attention to industrial requirements which had been largely ignored since the end of hostilities in 1918. In this sense, it was a highly significant point in British war planning.

Cabinet officially abandoned the Ten Year Rule on 23 March 1932. Once more though, finance was considered first and public opinion second. The

---

[31] *Ibid.*; and Gibbs, *Grand Strategy,* 78.

[32] It should be noted that the report, while mostly focussed on naval defences, also criticised the RAF and Army programmes for falling "far short" of the level needed for sufficiency.

[33] Gibbs, *Grand Strategy,* 80.

Cabinet meeting which formally abandoned the rule stressed that this did not mean increasing expenditure or abandoning the official policy of disarmament – still ongoing via conference in Geneva.[34] It could be seen as a tokenistic or meaningless gesture; the Ten Year Rule had remained a secret since its inception and the only outward sign of a change in policy would have been the simultaneous abandonment of the very public commitment to disarmament, which the government was not prepared to do.[35] On the other hand, in the minds of policymakers, if not the public, the hypotheses on which imperial defence were framed were now no longer based on the idea that there would be no major war for a decade, thus allowing progress to be made to tackle deficiencies.

Achieving this, however, was a major issue. Like the cruiser debates of the 1920s, the Treasury weighed in with a clear and persuasive case centred on financial limitations. It argued that even had the bases been fully reinforced some years before and thus could withstand attack until a fleet, modern or otherwise, could be readied in the Far East, Britain was in no position economically to sustain any sort of prolonged conflict. Thus, the question of improving the bases was moot until the economy had regained a sound footing. The Cabinet unsurprisingly also took this view. Furthermore, it believed that an agreement to reinforce bases would also mean stationing larger, more advanced warships in the Far East, and quite apart from the costs, the resulting increase in arms manufacture would also raise legitimate questions at the Geneva conference over the sincerity of British disarmament promises.[36] Importantly, however, MacDonald (albeit with more deliberation and delay) instructed the CID to prepare a report showing the "position of the private armaments industry in our system of Imperial Defence … today compared with pre-war times," comparing Britain to its foreign rivals.[37] Although this was a long way from a commitment to address the deficiencies, it was the first serious attempt post-Manchuria to understand them.[38]

Because of these delays, the official historian of rearmament has claimed that the National Government was guilty of "unbelievable tardiness" in understanding the implications of the warnings it was being presented with.[39] There is, however, another interpretation. By raising the usual problems of

---

[34] C.M. Bell, "Winston Churchill and the Ten-Year Rule," *Journal of Military History* 74/4 (2010), 527–528.

[35] Gordon, *British Sea Power,* 107.

[36] Gibbs, *Grand Strategy,* 81.

[37] TNA, CAB60/13, PSO359, "The Position of Private Armaments in British Defence," 10 March 1933. Note: PSO357, 6 March, has an identical title and refers to the same reports.

[38] For the purposes of this book, the third question on foreign competitors is not discussed detail, as it was sent to the Industrial Intelligence on Foreign Countries subcommittee for further analysis.

[39] Gibbs, *Grand Strategy,* 80.

finance and public opinion, while accepting the recommendations of the report in principle, the government shifted attention onto the remaining part of the conundrum: industry. Compared with the political and economic issues around rearming, the problem concerning the erosion of plant and skills was one where steps could be taken by the CID to examine the issues without the pitfalls of being publicly seen to be belligerent or financially irresponsible. In theory, this could have allowed a faster and more efficient rearmament in the future without attracting attention in the present. Such a move could not be done without some cooperation with industry, which did of course raise its own complications. Nevertheless, this was at least an area where there was some prospect of progress in the climate of the time.

Contingency Planning at the CID, 1932–1933

The CID had been diligently keeping abreast of Japanese developments since the late 1920s.[40] In this respect, the cancellation of the Ten Year Rule had given the work of subcommittees like the PSOC renewed meaning. Although the committee was formed in the early 1920s to coordinate the supply needs of the three forces, there was virtually no point before 1932 where its work had any real prospect of being immediately useful. Indeed, when the PSOC in 1929 had investigated the question of shipbuilding requirements in the first year of a future war, it quickly abandoned the project to focus on other tasks.[41] By the time the Prime Minister acted on the warnings from the CoS, the PSOC was already investigating the existing productive capacity available for defence.[42] As one member later put it, "the limiting factor was [no longer] the lack of something to investigate, but the lack of staff for investigations."[43]

Still, their work in the 1920s had given them something of a head-start. Indeed, when Maurice Hankey, the secretary of the CID, initiated proceedings to investigate the state of industry, he immediately circulated the two previous annual reports of the PSOC which had referenced the problem explicitly.[44] In other words, the problem was not with the PSOC's work but rather that Cabinet had

40 TNA, CAB60/35, PSO(SB)101, "Sixth Annual Report of the Principal Supply Officers Committee," 12 December 1929, appendix G.

41 This was not perhaps as significant as it appears – the PSOC busied itself with many loosely related and hypothetical issues in the 1920s without any long-term defence objective defined. TNA, CAB60/66, PSO(BT)42, "Labour Requirements," 10 January 1929.

42 TNA, CAB60/30, PSO(SB), 24th meeting, 11 October 1932.

43 TNA, CAB60/31, PSO(SB), 30th meeting, "Minutes of the Supply Board," 25 October 1933.

44 TNA, CAB60/13, PSO359, "The Position of Private Armaments in British Defence," 10 March 1933 (these were PSO318 and PSO350).

underused the organisation or taken little notice of its research for long periods. Manchuria was not so much a fundamental change in supply organisation – the structure remained the same – as a change in the Cabinet's attitude towards supply, which in turn gave the PSOC new purpose.

In the PSOC annual report for 1930/1931, produced before Japan had invaded Manchuria, private manufacturing capacity was already referred to as a "problem of peculiar magnitude and difficulty," with particular attention being drawn to the lack of "technical knowledge and trained personnel," and the issue of "expansion capacity [being] progressively reduced as peace time orders for munitions continue[d] to fall off."[45] In the following year's report, produced in the summer of 1932, special reference was made to the "contraction in the naval sphere of production capacity [as a] number of yards have been dismantled[46] and others are likely to be closed down." It also recognised that any "special forms of assistance" would require "political considerations outside the scope of this Committee."[47]

In response to MacDonald's request, the Controller of the Navy, Rear Admiral Forbes, produced a second up-to-date document detailing the Admiralty's estimated dependence on private firms in wartime. Again, this was based upon the hypothetical contingencies under which the PSOC was working, set by the CoS view of the most pressing threat to British security. Forbes pointed out that not all suppliers used were "armament" firms in the narrower sense, since "their existence in peace time does not entirely depend on the maintenance of orders for equipment." Nevertheless, the list gave cause for concern: ninety-five percent of hulls, ninety-eight percent of engines and one hundred percent of armour plate came from private firms, as did seventy percent of guns and one hundred percent of gun mountings. Moreover, although the Admiralty had only lost one firm from the pool of hull and engine manufacturers since the war, it had lost two of five from armour, two of four from guns and two of three from gun-mounting in the same time-frame.[48] Although not named, the two gun-making firms which collapsed in the period 1925–1933 were Beardmore and the Coventry Ordnance Works. In other words, for some components on which there had been total reliance on private firms in the previous war, the pool available to meet the same task in the next was in some regards much smaller than a decade beforehand.

45 TNA, CAB60/12, PSO318, "Extract from the Eighth Annual Report of the Principal Supply Officers Committee," 31 July 1931.

46 The term "dismantling" is important, since this is potentially more serious than a yard that has gone out of business. It was, in effect, a direct reference to the work of the NSS in sterilising capacity and removing the possibility, despite the merits of industrial rationalisation, of rapid reactivation in a national emergency.

47 TNA, CAB60/13, PSO50, "Extract from the Ninth Annual Report of the Principal Supply Officers Committee," 30 July 1932.

48 TNA, CAB60/13, PSO359, "Note and Statement by Third Sea Lord and Controller," 10 March 1933.

Moreover, it should be borne in mind that in 1914 Britain was in an extended naval race with Germany, which meant the firms involved possessed the latest equipment and technical knowledge. In 1933, not only were they far weaker in this regard, but the fleet was older and in more obvious need of modernisation. In many ways, therefore, the task was even greater than the numbers made out.

One further (and otherwise separate) investigation by the Subcommittee on Industrial Intelligence in Foreign Countries has content worth noting. This report showed that Italy now had as many ships and submarines under construction – thirty – as did Britain, and more importantly, had two orders for large and technologically complex ships to Britain's none. The situation had arisen primarily because Italian yards had succeeded in winning tenders ahead of British yards for numerous foreign naval contracts. Moreover, other rivals were becoming stronger. The Japanese Admiralty owned twelve factories, which were now not only self-supporting but due to enter the export market. This prompted the comment that there was "no doubt that the capacity of this country to build warships and produce munitions has been very substantially reduced since 1914" both in real and relative terms. The committee blamed the Washington and London treaties, the existence of "large stocks of equipment which are being used up but not replenished," and the lack of foreign orders for the decline. It concluded that the result had "gravely prejudice[d] the ability of this country to make warships" and had "dispersed the necessary technical staff."[49]

Perhaps the most important part of the report was its underlining of the fact that even the guns for the two large Washington Treaty battleships (HMS *Nelson* and HMS *Rodney*) had been delayed because of the lack of skilled men able to design and make gun mountings. So, even during relatively "lean" years for arms orders, forges were forced to recall pensioned members of staff to oversee and instruct the inexperienced younger generation who had never worked on such contracts before. It concluded by stating that, in marked contrast to Britain's own position, "no foreign country, with the exception of those defeated in the last war, has the capacity to manufacture munitions been reduced, compared with 1914."[50] In short, the two new papers – coupled with the earlier PSOC annual reports – validated Beatty's grim predictions more than a decade earlier.

[49] TNA, CAB60/13, PSO361, "Position of the Private Armaments Industry in Imperial Defence – Report by the Subcommittee on Industrial Intelligence in Foreign Countries," 8 March 1933, 7, 28 and 31.

[50] *Ibid.*, 7.

The Supply Board and Approach to Industry, 1932–1933: Virtual Unreality

Of course, pointing out the relative weaknesses that existed was only half the battle. In the absence of government will to place orders, there was still the job of finding other ways to improve industrial organisation should rearmament be required at short notice. The actual task of planning for this was undertaken by the Supply Board (see Appendix 2 for full membership; for its place in the CID structure, see Figure 1), which reported to the PSOC and comprised senior Air Force, War Office and Admiralty figures but had no representation in 1932 from industry or business. Since 1924, the Board's remit had been to consider broadly the potential problems of supply in Britain and the Dominions during a future war and to offer some solutions to likely issues. This planning body could not (and did not) fully eradicate these problems, as it had no budget and operated in secret, like the rest of the CID. The Supply Board lacked the ability, for example, to remove industrial bottlenecks through investment in the expansion of facilities. Instead, it existed to think through problems and to report upwards.[51]

The primary difference between the PSOC and its subordinate Supply Board was the latter's narrower focus on explicitly military items. The PSOC considered the problem from a broader perspective, taking in civilian supplies, contracts and anticipatory purchases. The Board's task was nevertheless still enormous in 1932: it coordinated and organised the provision of everything from entire warships down to medical supplies, which led to the creation of yet more and increasingly specialised subcommittees to consider the supply of each item or product. For example, Supply Committee (SC) I handled munitions, SC II focused on engineering, SC III was shipbuilding capacity and so on. While SC III was obviously of primary concern to the Admiralty, in other committees (particularly munitions) the three service arms interacted to discuss questions of supply and capacity together. It was thus the Supply Board's job to oversee and coordinate these diverse, but often competing, interests.

The Supply Board's history before Manchuria was much like its parent organisation: the geopolitical climate and existence of the Ten Year Rule had created a set of assumptions and the illusion of breathing space for solving issues. However, more than any other body the Supply Board was tasked with monitoring conditions in industry and looking for potential bottlenecks in production should an emergency occur. It was thus more attuned not only to developments in the established armaments industries for the three service departments but also to the related question of competition among them for the same pool of resources.[52] It also had an exceptionally senior membership. The Engineer in Chief of the Fleet (Harold Brown) and the Controller of the Navy (Charles Forbes) both sat on it,

---

51 Postan, *British War Production*, 35.

52 *Ibid.*

and they, along with the Director of Naval Construction (who did not sit on the Board), constituted the three most important figures in Admiralty procurement. The War Office and Air Ministry had similarly senior representation. It was therefore the single most significant committee regarding the defence industry, comprising some of the most knowledgeable men in the country at that time.[53]

Recent events closer to home had already influenced the committee's work. The wave of business failures from the mid-1920s to the early 1930s – the most notable of which were the giant armaments manufacturers William Beardmore of Dalmuir and the Coventry Ordnance Works – had been picked up by the Supply Board's antennae and reported along with the labour situation and the work of Sir James Lithgow's NSS scheme.[54] As such, it was more the pace rather than nature of work which changed. Taken together, 1932 was the year where systematic investigations into capacity got under way, but preparations for war were still a long way off.

The main objectives by 1932, although still strictly hypothetical, were to identify industrial capacity and allocate the productive resources of all the major armament manufacturers among the competing demands of the three service arms. This was achieved by allocating the capacity of a firm, for instance Fairfield, to the relevant Supply Board subcommittees – in this case SC III (Shipbuilding).[55] It was not the job of the Supply Board (or the PSOC) to hypothesise where or when war was most likely to break out; instead, it worked with the assumptions provided by the CoS.[56] These capacity investigations were thus built on a set of assumptions based on a shifting political outlook and economic base, and for that reason were undertaken as surreptitiously as possible. Enquiries were made on a discreet and informal basis as to what firms believed they could produce if working under emergency conditions.

From these investigations, a list of key firms was drawn up and allocated to the Admiralty, Air Force or War Office based on what their representatives felt was needed to meet its anticipated future requirements. For most firms, the allocation was relatively straightforward. The War Office or Air Force had no need, for instance, for most of the productive capacity of any of the naval shipyards, while the Admiralty did not want the services of the major rifle and small ammunition manufacturers. In some cases, the Supply Board thought that

---

53 Appendix 2 lists Supply Board membership.

54 TNA, CAB60/35, PSO(SB)53, "Rough Estimate of Quantities of Materials Required by Supply Committee III," 5 March 1929; and CAB60/66, PSO(BT)42, "Labour Requirements," 10 January 1929.

55 TNA, CAB60/30, PSO(SB), 24th meeting of the Supply Board, 11 October 1932, 12.

56 The PSOC had taken direction from the CoS on what to investigate since the 1920s, though the best example for how the system worked was the document, frequently updated, entitled "Basis on which Supply Preparations are to be Framed," which guided PSOC calculations. See, for example, Gordon, *British Sea Power,* 48; and TNA, SUPP3/18, CID1171-B, "Basis on which Supply Preparations are to be Framed," April 1935.

firms had sufficient capacity to supply more than one service, allowing their operations in the event of war to be divided by the production estimates given. Barr and Stroud, manufacturer of optical rangefinders, was a prominent and recurring example of a firm on the lists of both the Admiralty (for naval guns) and the War Office (for field artillery), while Messrs Saunders-Roe was split fifty-fifty between aircraft and boat manufacture.[57] In both cases the allocation was settled by agreement between the respective subcommittees.[58]

Of course, this supposed a certain kind of conflict. In the case of Barr and Stroud, the CoS believed in the early 1930s that any war would only require a small army and therefore a relatively small requirement for army rangefinders. In other words, while a certain hypothesis held, available capacity was sufficient. In cases where it was deemed insufficient, or where more than one service required the production facilities of a given firm, a different approach was needed. In one notable case, SC III (Shipbuilding), chaired by Engineer Vice-Admiral Harold Brown, "demanded the allocation of practically the whole of the Vickers works in Elswick on Tyneside to the Navy" for the manufacture of power-worked gun mountings.[59] The War Office also relied heavily in peacetime on the same plant for armoured vehicles, and this demand would certainly increase during a war. It therefore registered its own demands on the Vickers plant, while the Air Ministry did the same for fighters and bombers. It was the Supply Board's job in these circumstances to adjudicate.[60]

In the case of Vickers, Rear Admiral Forbes, the Third Sea Lord and Controller of the Navy, responded to Army and Air Force claims by arguing that "the provision of gun mountings had always been a nightmare for the Admiralty, as it was invariably the limiting factor in the construction of every kind of warship."[61] Pointing out that a large naval gun took at least twelve months to complete, he argued the Navy was especially anxious to ensure that this plant, along with Vickers' other gun factory at Barrow-in-Furness, was

---

57 TNA, CAB60/39, PSO(SB) 322, "Minute from Supply Committee VI," 20 February 1933.

58 TNA, CAB60/31, PSO(SB), 26th meeting of the Supply Committee, "Allocation of the Capacity of Messrs Saunders Roe between Supply Committees III and VI," minute 6.

59 TNA, CAB60/38, PSO(SB) 252, "Report by No. III Supply Committee," 1 October 1932.

60 TNA, CAB60/39, PSO(SB) 313, "Provisional Allocations to Supply Committee No. III," 17 February 1933.

61 The debate over gun mountings was, however, nothing new. Gun making was first recognised as the key factor in capital ship construction during the naval race with Germany before the First World War, and had remained an issue as naval guns became larger, heavier and more accurate over time. TNA, CAB60/30, 24th meeting of the Supply Board, 11 October 1932; and CAB 60/13, "Report by the Subcommittee on Industrial Intelligence in Foreign Countries," 13 March 1933. See also the report from the Foreign Industrial Intelligence Committee which argued along similar lines.

allocated to them, as "these alone were suitably equipped for this kind of work." Even if this was agreed, Forbes believed the situation with gun mountings was so dire that it would probably still be "necessary to augment these resources by placing some work with other firms," even if those firms were currently ill-equipped to fulfil orders.[62]

To help solve the problem, the members of the Supply Board agreed that Vickers had to be visited by both Admiralty and War Office staff to ascertain "within very fine limits" what could be produced in an emergency and how well Vickers' promises matched this figure.[63] As Vickers was under contract to produce equipment for the War Office and Air Force, and was a regular supplier to the Admiralty, the Supply Board had more concrete information about likely peacetime and emergency capacity at Elswick and Barrow than was the case for the majority of firms approached during these early phases. Even with this information, however, there was a degree of guesswork.[64] Furthermore, the Supply Board quickly realised that this method of allocating capacity was far from ideal; in a land war, it would clearly be beneficial to prioritise armoured vehicles first and naval guns later, while in a three-front war (land, sea and air) priorities would be somewhat harder to agree. As it was the job of the CoS to hypothesise who Britain was most likely to face in a war (and where), it was only for the Supply Board to ascertain how private industry would meet those demands. The Supply Board acknowledged that if a decision could not be reached, it would have to be referred to the CoS "with the view to the relative order or urgency from a strategic point of view being determined."[65]

In this instance, such a referral was not necessary. The visit to Vickers, led by Harold Brown and undertaken in late 1932, resulted in a decision to split the work of the Barrow and Elswick plants among the three defence departments, but protecting the Admiralty's first claim on facilities for manufacturing gun mountings. The visit was referred to afterwards by the Supply Board as "represent[ing] the first attempt to allocate the output of a firm between the three services, and a very valuable piece of constructive work." The same document indicated that the "method adopted in the solution of this problem might usefully be followed in the future."[66]

The Admiralty was not interested solely in the capacity of shipyards: by 1933 SC III had also visited steel foundries, optical munitions manufacturers and engineering plants. The subcommittee, for instance, put a claim in 1932 on most of the Weir plant for warship pumps, noting that it "relied upon" and always required around one-third of capacity, but "contemplated this being

62 TNA, CAB60/30, 24th meeting of the Supply Board, 11 October 1932, 13.

63 *Ibid.*, 14.

64 TNA, CAB60/13, "Report by Chairman of Supply Committee Nos. I, III and VI," 8 December 1932.

65 TNA, CAB60/30, 24th meeting of the Supply Board, 11 October 1932, 14.

66 TNA60/31, PSO(SB), 25th meeting of the Supply Board, 14 February 1933, 12.

doubled in an emergency." SC I (Armaments) felt that Weir's plant would be vital – as it had been in the last war – for shell machining, and it put its own claims on some of the facility. After another visit, the Supply Board found that Weir's was "a high class engineering firm" that was "well equipped" to make not only Admiralty pumps but also shells, gun carriages, gun mountings and aero engines, with the potential for rapid expansion. It concluded that while some balancing was necessary, SC I should be given fifteen percent of capacity for the time being, but that this could be reviewed if the plant was refitted for more munitions work.[67]

The decision to visit the Vickers and Weir plants did not mark any serious escalation or change of direction in war planning in 1932 and 1933 – there was no chance for the time being that Cabinet would authorise new orders for any extra equipment the defence departments wanted. Secondly, from a Supply Board perspective, making visits to plants was not common (although also not unique), but Vickers' and Weir's roles as major armaments manufacturers and regular and long-standing suppliers to all three service arms meant that visits would have been unlikely to draw any special attention or concern from the business community.[68] It does, however, highlight interesting deliberations over Britain's ability to produce key defence components. The problem with the lack of existing facilities to make guns and mountings had been severely exacerbated by the collapse of Beardmore in 1930, leaving, as Forbes noted, almost exclusive reliance on Vickers.[69] The naval gun issue should therefore not merely be taken as another case of Navy, Army and Air Force rivalry, a common feature of the 1920s and 1930s.

Similarly, this example does not provide any concrete allocations of capacity for future armament orders. Sir Harold Brown later said that much of the work done between 1932 and 1934 still had "an air of unreality" about it,[70] and during the discussion on guns the Supply Board made it abundantly clear that "all allocations, including this one, must be of a provisional nature" and that "approval of [the Vickers allocation] would not preclude members from raising the question again should further information … necessitate its reconsideration."[71] In other

67 GUAS, DC96/17, PSO(SCI)310, "Investigation of the Capacity of G&J Weir, Ltd.," 12 September 1932.

68 At least not in the way that visiting a factory ordinarily engaged in civilian work and never previously used as a defence contractor would be. In 1930, the SB noted that "owing to the ban of secrecy under which work is to continue, no abnormal enquiries must be made without ministerial approval." TNA, CAB60/30, PSO(SB), 12th meeting of the Supply Board, 18 December 1930, 13.

69 TNA, CAB60/30, 24th meeting of the Supply Board, 11 October 1932, 13; and Buxton and Johnston, *Battleship Builders,* 174.

70 BLH, TB, box 1, H. Brown, "Reminiscences," II, 110.

71 TNA, CAB60/31, PSO(SB), 25th meeting of the Supply Board, 20 December 1932, 13.

words, the agreement held as long as the calculations of the external threats to Britain remained the same *and* the general industrial picture remained unaltered. A shift in one or both would naturally affect or even invalidate any existing arrangements. Since both were complicated and fluid, these arrangements changed many times between 1932 and 1939.[72]

As far as the industrial picture was concerned however, the naval shipyards were struggling to find any orders at all: just 4000 tons of work was in hand for private manufacturers in 1931, including two destroyers for the Portuguese government, and almost ninety percent of berths lay idle.[73] Beardmore had closed, and Palmer teetered.

Approaching Industry, 1933

Along with discussions of firm's capacities, the Supply Board was debating another issue: "Local Area Organisation," a topic that brought members of the Board around to the idea of industrial cooperation in planning.[74] Unlike the supply question, which was in the main a response to the deficiencies highlighted by the CoS in early 1932, the revisiting of area organisation grew out of the experience of the Great War, where the sudden imposition of an organising body on top of firms' existing boards of management had led to "much friction and confusion," which the Supply Board believed "should in the future be avoided at all costs."[75] The desire for secrecy, however, meant that representatives from business were not called to give evidence before any part of the Supply Board or the PSOC. The topic was discussed only once before Manchuria, at the 17th meeting of the Supply Board in 1931, but the plan to consult industrialists did not get off the ground. Lieutenant General Charles felt then that "the time had not yet come to approach industry," but he did not rule out doing so in future should circumstances change. Percy Ashley, attending the meeting from the Board of Trade, "deprecated the [CID] at present of approaching any outside individuals," although like Charles he felt the question could be reopened in the future. The members therefore agreed that the point had not yet been reached when it was "necessary or desirable to make a general approach to leaders of industry and invite their cooperation in the work of the Supply Board."[76]

---

72 Brown, "Reminiscences," II, 112.

73 Warren, *Steel, Ships and Men,* 251; Johnman and Murphy, *British Shipbuilding and the State,* table 5; and A. Borthwick, *Yarrow and Company Limited, 1865–1977* (Glasgow, 1977), 54.

74 Brown, "Reminiscences," II, 110.

75 TNA, CAB60/31, 25th meeting of the Supply Board," 20 December 1932; TNA, SUPP3/18, PSO(SB) 298, Note by the Director of Navy Contracts, November 1932.

76 TNA, CAB60/30, PSO(SB), 17th meeting of the Supply Board, 28 August 1931, 9–10.

The circumstances in late 1932 were far removed from those of fifteen months earlier. By the 25th meeting in December 1932, some members believed that it was "essential that some organisation should be built up in times of peace [for organising regions] and that war time organisation should be based on this."[77] Beyond this general sentiment, though, there was still a gulf in opinion over how area organisation would look. Admiralty members of the Board believed that a system "based on the existing Admiralty scheme," where technical officers "stud[ied] each area in peace time and knew to where they would be despatched in an emergency, in order to form the nucleus of the area organisation" was the most desirable.[78] On the other hand, the Chairman, Air Marshal Dowding, had a "somewhat different conception" whereby the role of area organisations was not to organise the production of a special item or set of items, but to "comb out and exploit unknown local manufacturing resources for a variety" of different items, across the fighting services. Dowding cited the example of Siddeley's automotive works in Coventry, whose chairman had devised a scheme for war production that "made all other local firms subsidiary to his." Such an arrangement, Dowding argued, "would inevitably cut across the plans of other Supply Committees and endless confusion would result … [and] was a very great danger which must be provided against in peace time."[79]

For the most part these divisions reflected the nature of manufacturing for the Admiralty compared with the Air Ministry. The Admiralty was used to bespoke components being built by experienced suppliers (such as the members of the WSBC) in different areas of the country – the specific example cited by Vice Admiral Forbes was the experience of John Brown with steel and armour plate. The Air Ministry was more concerned with general suppliers, for example those in the textile industry, who could turn their hand to other simple mass-production techniques to help meet demand as it arose. Neither conception for the organisation of industry was better than the other; instead, it serves to highlight the problems arising from translating a system which may have worked for one defence department's suppliers into a rule generally applicable for all.

It must also be noted that any solution which allowed each department to devise its own system would certainly have had similar problems. Unless there was an agreed-upon standard, there was little to stop each from cutting across the bows of the others and undermining any attempt to share resources. It was as a direct result of this conundrum around regions that discussions began over alternative ways to organise industry. The contribution from Engineer Vice-Admiral Harold Brown in December, 1932 is worth quoting in full:

77 TNA, CAB60/31, 25th meeting of the Supply Board, 20 December 1932, 10.

78 *Ibid.*, 7.

79 *Ibid.*, 8.

> He understood that the Supply Board should utilise the heads of industry rather than the heads of specific firms. He understood that the authorities when studying the question in America had experienced difficulty as they had consulted heads of firms who had made plans for the production of their own particular product, irrespective of other requirements of the general productive capacity in their locality. He understood that they were now consulting people of influence with a general knowledge of their area [with] a view to assessing the most economic production.[80]

Brown went on to emphasise the importance of this question, indicating that in his view "the establishment of area organisations would be the means of implementing the mobilisation of industry." He also believed that "the sooner the Principal Supply Officers Committee spread its tentacles through the country, the better the task would be done and less likelihood would there be for a breakdown."[81] Jenkins, Charles, Forbes and Dowding all agreed, with the latter concluding that it was a "tremendous problem." The chair then instructed the three defence departments to jointly create the terms of reference for a special subcommittee to investigate the issue. This was the first, albeit very tentative, step towards working with industrial elites with a view to easing the problems of war supply in peacetime.

The Supply Board's remit did not allow it to make this decision without input from ministers. The Board frequently noted that any decisions which "involved questions of major policy, as well of finance, had to be referred to the Committee of Imperial Defence for formal approval."[82] Thus, in order be able to approach industry, the question first needed to be discussed higher up the chain of command, through the PSOC to the CID's executive committee chaired by the Prime Minister.[83] To this end, in March 1933 Walter Runciman, the PSOC's chairman and President of the Board of Trade, circulated the two earlier PSOC reports on the condition of the private armaments industry to CID members, including the Prime Minister, Ramsay MacDonald, the Foreign Secretary, John Simon, and the First Lord of the Admiralty, Bolton Eyres-Monsell. These were tabled at the CID's 258th meeting the following month.[84]

When the three men met as part of the larger committee on 6 April, MacDonald told the room that the reports made the situation appear "very serious" and that "it would be necessary to see what could be done from an industrial point of view to assist industry." He also suggested that "this matter

---

80 *Ibid.*

81 *Ibid.*

82 See, for example, TNA, CAB60/31, PSO(SB), 30th meeting of the Supply Board, 5.

83 Although the Prime Minister was notionally the head of the CID, much of the day-to-day running of the committee was undertaken by the CID's secretary, Maurice Hankey, who reported directly to him.

84 TNA, CAB2/5, CID1106, 258th meeting of the CID, 6 April 1933.

should be taken out of the hands of the Services and considered from a civic [sic] and industrial aspect."[85] Put another way, MacDonald had warmed to the view that business figures ought to assist in tackling the problem. Nonetheless, he was simply catching up with the view the PSOC and Supply Board had arrived at some months earlier. Despite Eyres-Monsell voicing concerns over the private naval shipbuilding industry declining in Britain while expanding in Italy and other nations, MacDonald refused to agree to any direct measures, citing his concerns about the possible effect on the disarmament talks in Geneva and efforts to control the international armaments trade. The meeting concluded with an agreement only to put in front of Cabinet proposals for yet more committees – without legislative power – to consider "remedying the serious situation that had been revealed."[86]

Cabinet approval for the PSOC to approach industry was received the following week. It also allowed for the formation of an inter-departmental committee to further examine defensive deficiencies. This body, the Defence Requirements Committee (DRC), produced its first report in 1934, and is discussed below.[87] Although nothing happened overnight, crucially there was at least movement within various parts of the CID organisation by mid-1933. Indeed, by the summer the topic had begun to appear on committee agendas under the heading of "method of approach to industry."[88]

Although moving slowly, there was clearly a paradigm shift in CID thinking away from the strict secrecy and isolation from industry that existed before the abandonment of the Ten Year Rule. At the 30th meeting of the Supply Board the discussion had advanced to a stage where Dowding felt that "immediate steps should now be taken" to "obtain the assistance of civilian experts in what, after all, was an industrial problem."[89] Others also urged "immediate consultation with industry," and at the following meeting in September 1933, the specifics of "taking advantage of industrial experience [to solve] the problem of locating capacity for war products" were finally discussed – but this was now nine months after some of the Supply Board's own members had first expressed support for the idea. Regardless, Lieutenant General Charles thought that in the first instance civilian experts should "advise on the general question of how work under the existing system might be accelerated" and hoped that "the assistance of industrialists might reveal sources of supply hitherto unknown to the defence departments."[90] Sir Reginald Townsend, attending as Director of Ordnance,

85 *Ibid.*

86 *Ibid.*

87 TNA, CAB16/109, DRC14, "Report of the DRC," 28 January 1934.

88 TNA, CAB60/31, PSO(SB), 30th meeting of the Supply Board, Minute 3: "Method of Approach to Industry," 2.

89 *Ibid.*

90 *Ibid.*, 2–3.

agreed and stated that he had found that "the advice of civilian experts was most helpful and always given willingly."[91]

Still, not everyone was enthusiastic. F.C. Bovenschen (from the War Office) expressed agreement in principle, but warned that it was "not clear at this stage what industrialists would be asked to do," while A.W. Bigsworth (from the Air Ministry) felt the organisation worked well in its current form. Others believed there were more pressing problems, including selecting a full-time chairman for SC I (Armaments), or were still doubtful whether any good would come of leading industrialists being brought into the fold.[92] Nevertheless, those in favour of approaching industry won the day: Chairman Dowding and R.K. Hezlett both firmly argued that it was "essential to approach leaders of Industry in the first instance," while Brown believed that the Supply Committee "required the assistance of industry in order to carry out its work." To this end, Charles prepared a memorandum which asked the PSOC to endorse proposals for the Supply Board to approach industrial experts in an advisory capacity, but to "leave executive action in the hands of the defence departments at present."[93]

On one hand, this was a triumph of sorts for the advocates of industrial collaboration given how reluctant the same committee had been only eighteen months earlier, or indeed how far off the prospect of receiving ministerial approval would have seemed in the aftermath of Labour's election victory. On the other, it does show a weakness in the committee structure, for despite many voices arguing for an approach to industry, there was still a long and painfully slow path to reaching any sort of decision on what was a rather basic point after some years of PSOC warnings about the state of Britain's productive capacity. Perhaps this was to be expected. After all, the CID and its subcommittees were designed to bring together and coordinate the competing claims of the three branches of the armed forces, and to reach compromises despite differing aims and objectives. As the investigations into naval capacity show, in some respects this was working well. Yet in a case where the Supply Board was expected to pick one choice out of a possible two options, the desire to reach a compromise merely meant months of delay.

The remaining question concerned which business figures to approach. Brown's views clearly carried weight, and Vice Admiral Forbes also believed that the "only way to get at the right men would be to approach the heads of the business" and "the big magnates" in the first instance. The terms used by Brown and Forbes, namely "heads of industry" and "big magnates" lack further definition in the records of the committee. After all, the most senior figures in industry were also typically owners and chairmen of firms and thus magnates.

---

91 *Ibid.*, 4.

92 *Ibid.*, 5, 7.

93 TNA, CAB60/40, PSO(SB)376, "Memorandum by Master General of the Ordnance, War Office," August 1933.

However, one can suggest – and this is borne out by the men eventually chosen – that Brown and Forbes were thinking particularly of the relatively small number of men who had transcended the business sphere into public and political life as spokesmen for, and well known figures in, their respective industries. Cobb, the Deputy Director of Navy Contracts, was by this point the lone dissenting voice, believing that "in many cases leading industrialists were merely financiers with little knowledge of technical or detailed working," and suggesting an approach to subordinates instead to get them to work with the individual Supply Committees rather than the Supply Board or PSOC.[94] Dowding agreed with Brown and Forbes, pointing to the "invaluable assistance" that Sir Harry McGowan, Chairman of Imperial Chemical Industries (ICI), had once provided regarding explosives through loaning the technical expertise of his staff.[95] In addition, letters from Weir and Thornycroft to the *Times* were circulated "as a matter of interest."[96] In retrospect, it perhaps should not come as a surprise that the "heads of industry" view won out. After all, the Supply Board's main concern was with the organisation and coordination of industry, as well as maintaining secrecy as far as practicable. Thus, approaching a small coterie of businessmen who had risen to the very top of the tree served the board's remit better than subordinates with technical knowledge of more specific manufacturing processes.

It was thus agreed that the Controller of the Navy, the Master General of the Ordnance, Browett from the Board of Trade, and the Chairman should consider which suitable individuals should be consulted, and that these names should be laid before the PSOC "without delay." It was also deemed highly desirable to have a meeting with these industrialists and the Chairman of SC I (whose problems were believed the most urgent) before the end of the 1933 "to acquaint them with the system of supply and discuss general principles." In addition, they should "be given an idea of the organisation and the hypotheses on which work was being conducted."[97]

In sum, the agreement reached at these meetings represented a major milestone and a peacetime first. Though progress had been painfully slow and the process inefficient, in the end the formal foundation of a panel of civilian industry advisors within the CID had been agreed. This was no mean feat.

---

94 TNA, CAB60/31, PSO(SB), 30th meeting of the Supply Board, 25 October 1933, 4–5.

95 McGowan's involvement in 1933 was far less than the proposed collaboration with industry, however. While Dowding clearly appreciated McGowan's assistance, he was only being asked the opinion of his company on specific technical questions and was not involved in matters of supply organisation.

96 TNA, CAB60/40, PSO(SB)387, "Industrial Mobilisation: Note of Letters to *The Times*," 30 November 1933.

97 TNA, CAB60/31, PSO(SB), 31st meeting of the Supply Board, 9 February 1934, 7.

Chapter Six

The Advisory Panel of Industrialists and Inside Information, 1933–1934

In the event, the men picked were not "merely financiers:" they were Lord Weir, Sir James Lithgow and Sir Arthur Balfour, three extremely powerful businessmen from engineering, shipbuilding and steel, respectively. Weir's background has already been discussed in the context of the PSOC, but by this point had further established G&J Weir Ltd. of Glasgow as a truly global enterprise. Weir's firm had also turned rapidly and with great effect from civilian production to war materiel, including boiler pumps for naval vessels in the Great War. As one of the architects of the PSOC structure that he was now being asked to assist, Weir was held in extremely high regard for blending business acumen and leadership with technical knowledge, distinguishing him as a rare breed. During the latter stages of the Great War he had served for eight months as Lord President of the Air Council and had subsequently been chosen to serve on the committee which investigated the role of the Fleet Air Arm in 1921. Arguably his most important contribution had been his work in the Ministry of Munitions in the Great War, and it was this that underscored his credentials for the PSOC. Weir had since cemented close connections with high-ranking members of the Conservative party; after serving on the Salisbury committee in 1923, he kept frequent private correspondence with, among others, Stanley Baldwin and Neville Chamberlain – the latter referring to him in an unusually friendly manner in letters as "My Dear Willie" – both of whom stayed at his Eastwood Estate in Glasgow on several occasions.[1] He was extraordinarily well respected, well connected, and an obvious choice to assist the Supply Board.

Colonel Sir James Lithgow has also been discussed in earlier chapters. Like his friend and mentor Weir, he was a successful business owner with a track record of government service during the war; Lithgow served both at the front and as the Director of Merchant Shipping. Furthermore, although he had no experience in armaments manufacture – his Port Glasgow firm was a leading provider of merchant and passenger vessels – he had founded and led the NSS scheme since 1930 and served for three years as the President of the Federation

1 GUAS, DC96/22/2, Chamberlain to Weir.

of British Industry.[2] Lithgows Ltd. also had coal and steel interests, and Sir James was heavily involved in the rationalisation of the Scottish steel industry. Thus, like Weir, he was chosen for his role in overseeing and coordinating industry, and for the unusual level of experience he could bring to the question of supply organisation.

The third member of the panel was Sir Arthur Balfour, chairman of the Arthur Balfour & Co. (formerly Seebohm and Dieckstahl) steelworks in Sheffield. Balfour has not appeared in the narrative before now, but he was of a stature comparable to Weir and Lithgow. His firm manufactured steel and armour for all three service arms, and, like his counterparts, he had long-standing relations with the state. In the Great War, he was a member of both the Advisory Committee on War Munitions and the Industry Advisory Committee to the Treasury. In the immediate aftermath, he was one of three independent members on the Coal Industry Commission, a major enquiry set up by an Act of Parliament as a Royal Commission to investigate the state of British coal mining, with the stated intention of improving productivity. Although then-Prime Minister David Lloyd George did not act on the Commission's findings, Balfour's work gained him great respect, and he was knighted in 1923 before becoming President of the British Chambers of Commerce and later serving on both the Committee on Industry and Trade as well as the Economic Advisory Council between 1924 and 1930.[3] He complemented Weir and Lithgow, and brought experience in raw materials and productivity questions in national industries such as coal and steel – a desirable skill-set.

None of the industrial members selected were from firms represented in the WSBC. This may appear strange, given long-standing relationships of some of these firms with the Admiralty and the importance of firms such as John Brown and Vickers to the war effort before 1918. A few reasons may be suggested for why men such as Charles Craven (managing director of Vickers) were not included in the Advisory Panel. The first is that Weir, Lithgow and Balfour all had long and proven track records of working with the government, not just fulfilling contracts for the armed services. All three had records of *public* service that had brought them into close contact with the top echelons of government (to which the CID and PSOC ultimately reported), which should not be underestimated. When Weir made his opinions known, as he did to *The Times* to support Thornycroft, the CID took note.[4] Perhaps most revealingly, a 1933 memorandum circulated to the CID (and later shared with the Advisory Panel) by G.S. Witham, the Assistant Director of Ordnance Factories at the War Office, noted that "many of the men who had

---

2 Reid, *James Lithgow,* 90–91, 132, 193–203.

3 Mary Walton, "Balfour, Arthur, First Baron Riverdale (1873–1957)" *Oxford Dictionary of National Biography* (Oxford, 2004); online edition, October 2006 [http://www.oxforddnb.com/view/article/30552, accessed 9 October 2013].

4 TNA, CAB60/40, PSO(SB)387, "Industrial Mobilisation: Note of Letters to *The Times*," 30 November 1933.

actual experiences of war production in 1914–18 have passed out of industry. It is therefore a matter of considerable urgency that those who remain in industry should be harnessed into the organisation."[5] Weir, Lithgow and Balfour certainly fitted the bill in this regard.

Secondly, concern over the sensitivity of information was a recurring theme of Supply Board discussions, and it was therefore determined that the numbers of industrial advisors be kept as small as possible. The first discussions mentioned a panel of "no more than six," which was then cut to four, and finally to three. In this regard, the three who were chosen had plentiful experience – and had earned a degree of trust – in dealing with secret information of this kind. As all had risen through their firms and faced the challenges of the Great War, they were among a small group with both technical and managerial experience. Craven, despite his role as managing director of Britain's largest armaments maker, Vickers, had not reached this level by 1933. He had not served in wartime ministries (he was a twenty-nine-year-old submarine lieutenant in 1914); he had not served on post-war commissions; his experience leading Vickers had been exclusively in peacetime (from 1924); and he had not yet been knighted for his public service. While all of this would come in due course (he did play a significant role in the Second World War), his *vitae* did not match those of Weir, Lithgow or Balfour.[6] Finally, the fact that the trio were not explicitly armaments manufacturers (even if their wares – at least in the cases of Weir and Balfour – were bought by the defence departments) probably helped. Their remit concerned supply organisation for industry and thus any interest that aligned too closely with one set of defence priorities may have been undesirable.

The Advisory Panel and Information Sharing: A Two-way Relationship

The three men were invited to join in October 1933 and would be known and listed as the PSOC's "Advisory Panel of Industrialists" until 1939. This invitation should also be viewed through the prism of concurrent events, most notably the ongoing NSS sterilisation efforts. Deliberations on who to approach in industry had, after all, taken place in a fluid period. Compared with 1932, the process of liquidation of warship yards appeared, if anything, to be increasing: Palmer was in the process of being sold to the NSS, and parts of Fairfield or Armstrong were looking very much like the next in line. Thus, Bigsworth (Air Ministry), Charles

---

5 GUAS, DC35/31, Witham Memo, February 1933 (cited in Lithgow papers).

6 Moreover, Vickers' Chairman, Herbert Lawrence, was much more in the mould of the "financier" that Cobb described, with a background almost wholly in banking. Lawrence only joined the firm as director in the mid-1920s, and by 1933 was seventy-two-years old.

(War Office) and Forbes (Admiralty) from the Supply Board, plus representatives from the Board of Trade, formally met Lithgow, Weir and Balfour in their new capacity as advisors for the first time only a few weeks later at Whitehall on 19 December. The President of the Board of Trade, Walter Runciman, chaired the meeting, which was arranged to "explain the present organisation for, and the situation as to, industrial mobilisation, and to discuss the acceleration of the turnover of manufacture from peace to war." Despite the collaboration being in its earliest stages, the panel had been supplied in advance with some details of how the CID and its subcommittees operated, and so were invited (after a short re-cap by Charles) to comment on the structure before being asked to take on the task of preparing a report on capacity and efficiency within industry.[7]

Interestingly, the responses appear to indicate that they still had only a limited knowledge of the work of the CID, the PSOC and the Supply Board. Weir, who had by far the most prior knowledge of the CID, commented that the "[Supply Committee] organisation, which presumably represents the central policy of His Majesty's government, appeared adequate," while Balfour was "pleased to see that a complete organisation was in existence."[8] Lithgow required both explanation and clarification on the matters of "secrecy [and] the methods by which firms were approached prior to the investigation of their works," which he was provided with in January 1934.[9] At this stage, the industrialists could only make general comments, but all broadly agreed that the situation could be improved and that industry was, in many ways, weaker than in 1914. It was then agreed to furnish the panel with the reports of the PSOC from 1932 and 1933, as well as "memorandums and a chart dealing with the Organisation at present in existence."[10]

Gaps in knowledge of this nature – even in the minds of men with extensive experience of wartime industry – should not be surprising. After all, the work of the PSOC's Supply Board was highly secret and quite unlike the previous Ministry of Munitions, so for Lithgow to know little of the extent of the individual Supply Committees' investigations into capacity is perhaps to be expected. It demonstrates the success of keeping such an organisation out of the public domain to the extent that these "really big magnates" were only having the basics of supply organisation fleshed out to them in December 1933. The members of the panel were also given relatively little extra assistance to support them with their work – Reginald Townsend was primarily concerned about the PSOC finding "adequate staff to keep pace with the recommendations likely to result."[11] This in part highlights the fact that the CID, despite its numerous subcommittees,

---

7 TNA, CAB60/14, PSO 407, "Cooperation with Industry," 19 December 1933.

8 *Ibid.*, 3.

9 GUAS, DC35/31, Witham to Lithgow, "Summary of Firms Visited, 1930–1933" and "Some Notes on Process Specifications," 2 January 1934.

10 TNA, CAB60/14, PSO 407, "Cooperation with Industry," 19 December 1933, 4 and 5.

11 TNA, CAB60/31, PSO(SB), 30th meeting of the Supply Board, 25 October 1933.

was understaffed and overworked, especially given the importance of its tasks. Moreover, it also shows how much the industrialists had to acquaint themselves with in a short space of time.

The panel members were provided with more secret background material throughout January and February of 1934. Assisted by the Supply Board's joint-secretary, Lieutenant Colonel Hind, and Witham, the material included reports from the Admiralty on procurement, reports from the Master General of the Ordnance on the allocation of experimental and developmental work, and a memorandum from SC I on the production of armaments in an emergency. The information was highly revelatory. Lieutenant General Charles, the Master General of the Ordnance, noted the "dwindling choice of firms capable of manufacturing stores which are non-commercial in character." In addition, Hind provided Lithgow, Balfour and Weir with a twenty-page document detailing all the "key" firms visited in the past three years, the capacity currently available, and what would be needed in the event of a war. Beardmore, for instance, would take five months to make a single gun-carriage for the War Office, but within eight months could make eight carriages per week.[12]

Perhaps most interestingly of all, the Director of Navy Contracts, Jenkins, wrote in his report that the Admiralty:

> … varied the policy of inviting competitive tenders in the case of certain armament firms whose designing staff and productive capacity is [sic] essential to maintain as a nucleus in an emergency. In such instances, it is customary to confine the orders to these firms on condition that fair prices are charged and that they can meet delivery requirements.[13]

In other words, the Director of Navy Contracts was confirming what the WSBC believed but did not conclusively know: that the Admiralty was primarily interested in spreading orders around a group of specialist facilities if the price appeared fair (and, one presumes, they accepted the lowest tender from the invited firms). Jenkins also drew attention to the Admiralty's difficulty with "providing enough work in peace time for these 'key firms.'"[14] As this report was written in March 1932, Jenkins almost certainly never explicitly intended for someone like Lithgow, a shipbuilder, to read it. Events had clearly overtaken these plans by 1934, however, and for a businessman to have this sort of officially secret information – and in writing no less – was both highly unusual and reflected the unique set of circumstances facing the CID and its supply subcommittees at this point.

---

12 GUAS, DC35/31, PSO(SB)39, Eighth Annual Report of the PSOC, 1931, SC I, "Proposals for Acceleration of Progress," undated.

13 GUAS, DC35/31, PSO(SB)226. "Report from Director of Navy Contracts on Commodore Bigsworth's Memo," 22 March 1932.

14 *Ibid.*

Recommendations

To their credit, the panel's members wasted no time in undertaking their task. By the end of February, just nine weeks after the first meeting and two weeks after they were furnished with the last round of further information, Lithgow, Weir and Balfour had produced and circulated a report (often incorrectly referred to as just "Lord Weir's Memorandum") outlining their views on the basic problems with the organisation of industry, as well as providing some suggestions for improving the existing structure. The panel began by apologising for the "informal nature" of the report owing to the speed with which it was prepared, which had meant that they had not been able to "intensely study" the full question. Thus, the report indicated that it would be confined to the examination of relatively simple objects, namely the production of shells and munitions (the responsibility of SC I) rather than the engineering, armour and shipbuilding sectors of which the panel's members had more working knowledge. In addition, the panel had some clear misgivings about suggesting ways for the state to reorganise or interfere with normal business practices, and it put on record the "personal distaste with which [they] regard[ed] the work." Indeed, they were at pains to note that they justified it to themselves "on the fundamental basis of the importance of national security alone."[15]

Nevertheless, Lithgow, Weir and Balfour[16] were not short of suggestions. They first recorded their fears that the "coordination of supply will require something much more effective than the current structure in the event of a [world war]," something they viewed as only adequate for peacetime or limited conflicts. They also believed that the peacetime industries (such as shipbuilding and engineering) were better equipped to prepare for war than those needed for shell and munitions production, as the latter were only needed in substantial quantities in the event of a large-scale conflict, making an expansion of industry during the peace highly problematic. However, they insisted that this did not mean that shipbuilding and other areas required less attention and warned against "frittering away the valuable opportunity" to "give definite support" to a small number of units. In other words, like the Admiralty they wanted orders to be distributed among the larger armament manufacturers to build up their capacity and expertise in preparation for a period of high demand. Failing to do so, they argued, would only "maintain in existence ... a series of weak and inefficient units unable to deal with an emergency or to act as progressive units in time of war." This was particularly serious, for during wartime such units should be "the most valuable assets for armament supply ... due to their peacetime scale

---

15 TNA, CAB60/14, PSO 415, "Memorandum by Lord Weir, Sir James Lithgow and Sir Arthur Balfour," 20 February 1934, 1.

16 Weir and Lithgow co-wrote the report, as Balfour was in America. The latter did, however, sign off on the contents of the document, and added a three-page postscript.

of output" and thus had to be "able to carry an administrative and executive structure, strong in organising and technical personnel coupled with toolroom strength, metallurgical knowledge and experimental facilities."[17]

More generally, the report indicated clearly the industrialists' convictions about the limits of planning. They note that while weaknesses can "be met by planning and prevision," they were "convinced that however perfect [Britain's] organisation may be, no mere paper planning can cope adequately with the foreign situation" which, in their eyes, would have Britain in conflict with nations that possessed a much "strong[er] peacetime armaments industry." Moreover, they expressed concern that "all technical, design and production facilities" for all types of weapons (including warships) were dependent on a tiny number of firms or state bodies. In the case of SC I (Munitions), production relied completely on Vickers and the state-owned Woolwich Arsenal, whereas in 1914 there were six comparable facilities. Weir, Lithgow and Balfour noted that "in these two units reside all the technical data and experience of this country," and should provide a kind of "parental guidance" to less experienced firms. They were also not convinced that Woolwich in South East London was safe in the event of aerial attack.[18]

The panel's solution was the dissemination of industrial knowledge among some of the key firms which had been identified over the previous two years. The reports given to the industrialists contained both the listing of the thousands of firms previously been engaged in armament work of any kind and the extent of the investigations (such as that at Vickers) which had been undertaken at dozens of facilities around the country. The Advisory Panel deemed the first list as "too broad" to work, and instead recommended investigation of a much smaller number – 250 to 400 – of "major" firms with "all attention being concentrated on them" for the mobilisation of industry for all types of armaments from rifles to ships and aircraft. These firms, they suggested, would almost all be well known to the War Office, Air Ministry and Admiralty, meaning more accurate estimation of their skills and theoretical capacities.[19] Lord Weir had previously offered to "advise as to which firms were most suitable," but he warned that approaches should only "be made to the big firms and the big men."[20]

Essentially, what the panel advocated was a system of what would be most frequently known as "educational orders," whereby an already trusted firm could develop skills that it did not possess in peacetime to be ready to ramp up production of items needed during war. The items made during peacetime would initially be expensive and slow to manufacture, but it would allow for an invaluable knowledge base to be built up while simultaneously adding to

[17] TNA, CAB60/14, PSO 415, "Memorandum by Lord Weir," quotations 2–4.

[18] *Ibid.*, quotations 4–6.

[19] *Ibid.*, 6.

[20] TNA, CAB60/14, PSO 407, meeting, 19 December 1933, 5.

Britain's existing stock of weapons and munitions. This "shadow" industry could then be placed on a wartime footing quickly should the need arise. In the case of naval shipbuilding, where the number of berths were adequate but skills and facilities had eroded over time, educational orders would enable a new generation of apprentices to be trained and would allow much shorter lead-in times between when a ship was ordered and when it was finally commissioned for service. Such ideas were not new – Weir had said the same thing several times since 1931, most recently in a letter to the *Times*[21] a month before meeting the CID – but in February 1934 the audience was very different. Moreover, it cannot be ignored that these proposals also served the interests of the panel members, who were all active in firms that would inevitably benefit from the policy.

As an intriguing side-note to the recommendations, the report also sheds light on the limits of the information shared with the industrialists in early 1934. For instance, the beginning of the report noted an "underlying reservation" that they "do not know and cannot question the hypothesis on which our war needs are based. When a figure is mentioned, we accept it." In addition, when the figure of twenty million shells for the first six months of a war was discussed, they noted that "without further information, it is impossible to be dogmatic on whether [the existing facilities] make a sufficiently strong foundation for this country."[22] Moreover, the limitations on the information available to Weir, Lithgow and Balfour were not confined to just the subject of defence hypotheses. While this may have been deliberate, it is perhaps more likely another reflection of how little time and resources were available to allow for a fuller knowledge of the CID. At the 32nd meeting of the Supply Board, two weeks after the industrialists' memorandum was submitted, the Chairman, Air Marshal Dowding, remarked that on the subject of supply organisation, "Lord Weir had not been given full information about the whole process," while Lieutenant-General Charles said that the panel's proposals for spreading orders around more firms were perhaps made because they "did not realise how minute the present Service orders were."[23] Admittedly, the panel had only been working for the PSOC for a matter of weeks, so it is perhaps unrealistic for these hypotheses and processes to have been shared from the outset, but it is at least clear that while the industrialists had been given an unprecedented amount of secret information, there was still a great deal that was unknown to them.

The report was, nevertheless, taken very seriously by the Supply Board. The members were so impressed that they agreed that the industrialists should be brought back in as soon as possible and allowed to elaborate on their proposals

[21] Lord Weir, "British Shipyards, Importance of Varied Work, Continuity of Technical Progress," *The Times*, 17 November 1933.

[22] TNA, CAB60/14, PSO 415, "Memorandum by Lord Weir," 3–4.

[23] TNA, CAB60/31, PSO(SB), 32nd meeting of the Supply Board, 21 March 1934, 15.

in front of the PSOC.[24] This did not mean that the PSOC and the Supply Board were now prepared to invite industrialists into the fold in larger numbers; indeed, during the 33rd meeting in May, the PSOC's members reiterated their belief that while industrial cooperation was now essential, adding more civilians as permanent members was not wise. Witham noted that "their presence might be found embarrassing, and might reasonably offend the susceptibilities of their trade competitors" if the information became widely known. Jenkins stated that his long experience of employers' groups led him to think that their help would not be very satisfactory, as they were always apt to safeguard the interests of their least efficient member. In his opinion, it would be more profitable to get in contact with an outstanding individual in a trade and take him into full confidence only if he fitted the bill, in the same way they had done for Weir, Lithgow and Balfour. They did, however, conclude that this was only for the present and that they would consider changing it in future.[25]

The meeting with the Advisory Panel took place on 9 April, and culminated with Dowding producing a lengthy memorandum to respond to each of the individual points raised by the panel and to provide its members with some more information on the state of procurement. Some aspects of this memorandum are especially noteworthy. About the decline of the armaments industry, Dowding wrote that the:

> ... gravity of the situation, revealed by the weak state of the armament industry in Great Britain, has long been fully realised. It was brought to the notice of the [CID] in the Ninth Annual Report of the PSOC (in 1932). The Supply Board are very conscious of the extent to which the field of armament supply has dwindled and concur wholeheartedly with Lord Weir's proposal to create a shadow armament industry capable of meeting at least the major part of our war requirements. Peace requirements are at present so small that they are almost fully absorbed in maintaining capacity in Government factories ... The acceptance of this policy, which would be of the greatest value in widening the field of supply, would therefore entail some increase in peace-time orders and possibly also some increase in the cost of peace-time contracts. It is considered, however, that the increase of war time capacity thus achieved would amply compensate for these disadvantages.[26]

In addition, he also gave the PSOC's endorsement to the concept of educational orders to some non-armament firms and to an immediate selection of 400 of

---

24 *Ibid.*, 16.

25 *Ibid.*, 12.

26 TNA, CAB60/14, PSO 421, "Note by the Chairman of the Supply Board on the Memorandum by Lord Weir," 26 April 1934, 2–3.

the most suitable firms for War Office, Admiralty and Air Ministry contracts that already possessed a "nucleus of plant and experienced personnel." He also reaffirmed his commitment to a small advisory body and that more experts "should not, at present, be brought into active membership of any of our subcommittees." Responding to Weir, Balfour and Lithgow's suggestions for the mobilisation of current armament manufacturers, Dowding reported that he had informed the panel that the PSOC was already, in fact, making progress on this matter, and that schemes regarding the "capacity of Messrs Vickers-Armstrong and Beardmore are under preparation in conjunction with these firms."[27] Moreover, on the issue of shell filling, the panel were informed that a Great War facility in Hereford had been kept on a "care and maintenance basis" so that it could be turned over to production relatively quickly should circumstances dictate.[28] Given Lithgow's involvement by 1935 with Beardmore and his NSS scheme, these pieces of information on top of that already shared with the panel would become very significant indeed.

The PSOC, and indeed the wider British government, benefitted immediately from the cooperation of this small group of industrial advisors. While in the area of naval arms the industrialists were not suggesting any radical overhaul to existing practices, the "shadow" scheme was proposed to reduce the future strain that would likely be placed on firms like Vickers, allowing it to expend its efforts in other areas. The same could be said of educational orders, which were motivated by the desire to reduce the problem of securing a steady supply of items that only a handful of firms could at present fulfil. The suggestion to focus any effort on a much smaller pool of firms than had hitherto been the case, self-serving as it may have been, was perhaps just as important because it allowed the work of a small committee to be more effectively directed towards hundreds, rather than many thousands, of facilities.[29]

Of course, the industrialists were also gaining from their new relationship with the CID. As businessmen whose firms had long benefitted from the British government contracts, access to secret information on supply organisation and likely procurement bottlenecks (in other words, what a major customer might need soon) had obvious value. But regardless of the panel's motivations or the recommendations that Dowding and the Supply Board may have been endorsing, the Supply Board and PSOC were still in no position to act upon them in April 1934. Still, this was a crucial first step towards understanding where the problems in supply might be and developing a dialogue to overcome them. It was also progress towards what would eventually be implemented following the

---

27 *Ibid.*, 5.

28 *Ibid.*

29 Before the decision was made to create a panel Bovenschen conceded that a major stumbling block was not a "lack of something to investigate" but a "lack of staff" and knowing "which firms to go to first." TNA, CAB60/31, PSO(SB), 30th meeting of the Supply Board, 25 October 1933.

publication of the first White Paper on Defence – which brought rearmament into public consciousness – in March 1935. The financial and political situation at home and abroad in mid-1934 meant that this remained, like most of the PSOC's plans, a hypothesis. In retrospect, it is unlikely that this scenario of cooperation and extensive planning would have arisen at all in the wake of Manchuria if there had been no National Government. The "spark" was the revocation of the Ten Year Rule, which required Cabinet support, and it is highly doubtful whether the Labour Cabinet of 1931 would have backed it in the same way. The unique blend of extreme financial and political pressures necessitated an approach which planned much but spent little, and this is one important reason for the formation of the Advisory Panel rather than skipping straight to a form of limited rearmament in 1932.

Related Business Developments: Lithgow and the NSS, 1933–1934

The implementation of these recommendations is discussed below. At this stage, though, attention must return once more to Lithgow's NSS scheme. As suggested earlier, the NSS can be seen in the context of the pre-Manchuria period as another in a long line of attempts by industry to cope with the dire market conditions of the 1920s, although for our purposes the history of the NSS after 1933 is best understood when placed in the context of the CID's changing defence hypotheses and the subsequent approach to industry.

From its inception in 1930, the NSS had successfully liquidated hundreds of thousands of tons of shipyard capacity. For the most part, this was not of concern to the PSOC, as only a relatively small part of merchant capacity was used for naval construction during the Great War, and there were still many hundreds of berths left. Beardmore was a larger issue because, along with Vickers, it was responsible for making components of naval guns and other complex products. Its closure had therefore left a gap that, by 1934, *was* a major cause for concern. Fortunately for the PSOC, Beardmore was split over two sites in the west and east of Glasgow, namely the shipyard in Dalmuir and the forge in Parkhead. The closure of the shipyard under the NSS scheme had naturally been concerned with rationalising shipbuilding capacity, and by April 1933 work had been completed to liquidate and sell off the yard's machinery. The NSS did not purchase all the firm's engineering plant or its forge – these were outside of the terms of reference of Lithgow's scheme, and as such remained under the ownership of the Bank of England.[30]

Incidentally, one of the outcomes of this arrangement was that Lithgow – who refused an earlier offer from Norman and Duncan to become chairman of

30 Hume and Moss, *Beardmore*, 216, 217.

the firm when the Bank initially invested in Beardmore in 1928 – agreed to sit on the board of the surviving part of the firm, with the intention of advising the company on rationalisation (and, one assumes, being a conduit back to the Bank).[31] This should not be overstated at this juncture because Lithgow played only a minor role before 1935, and there is no indication that he knew anything about the WSBC. In sum, although Beardmore's capacity to make most aspects of warships and naval guns had been eliminated, its existence as an engineering and steel company meant that the rump of the business survived, and some naval arms production capacity – especially in armour plate – remained.[32]

Until 1933, Lithgow and the board of the NSS had favoured the liquidation of entire yards or companies, rather than individual sites or berths belonging to a larger shipbuilding firm.[33] In this respect, the diversity and size of the Beardmore business made it an unusual case. At the outset, this policy of large-scale liquidation caused relatively few issues as there was no shortage of sellers, but after more than two years of sterilisation the plan began to run into trouble. Beardmore was by this point one of only a relatively small number of yards capable of making warships which had been liquidated by the NSS – the warship and liner sector was proving harder to liquidate than the merchant sector. The main sticking point was that those constructing the simpler cargo vessels could be bought wholesale, while the more specialised yards held some assets which their owners remained unwilling to sell, even if many of the firm's berths were unoccupied, or where they were willing to sell off parts of the business. Thus, by the conclusion of 1932, progress with merchant berths had been far quicker than the equivalent warship and liner berths, which were almost always parts of much larger and more technologically advanced yards, and concentrated in the hands of a few firms.[34] Thus, an urgent meeting was called among the directors of the NSS and the major warship manufacturers to discuss the problem in early 1933.[35]

The meeting, chaired by Lithgow, took place on 23 February, and naturally included most WBSC members. Their representatives included Sir Thomas Bell and Captain T.E. Crease (of John Brown), W.L. Hickens and R.S. Johnson (of Cammell Laird), Sir Maurice Denny (Denny), Sir Alexander Kennedy (Fairfield), Sir Frederick Rebbeck (Harland & Wolff), Captain G. Bates (Hawthorn Leslie), Sir John Scott III and James Brown (Scott), A.M. Stephen and A.L. MacLellan

---

31 Murphy, "Déjà Vu all over again," 64.

32 GUAS, GD320/2/10, William Beardmore, Directors Report, 16 April 1934; and Hume and Moss, *Beardmore,* 218. The survival of the engineering plant is critical – as some of the facilities for gun making were preserved.

33 NMM, NSS, minutes, various locations, but see especially Barr's input from 16 December 1932.

34 See NMM, NSS, minutes, 16 December 1932. For an example of the notice sent to the warshipbuilders, see GUAS, GD320/2/10, Lithgow to Scott, 23 January 1933.

35 GUAS, GD319/12/7/6, Lithgow to Scott, 27 January 1933.

(Stephen), Sir H.A. Lawrence, Colonel Beaumont Nelson and Charles Craven (Vickers Armstrong), Sir George Goodwin (Samuel White), and Sir Harold Yarrow (Yarrow). Aside from the defunct Beardmore and troubled Palmer companies, only Swan Hunter and the destroyer builder Thornycroft did not attend the meeting. Here, the odd scenario arose where the NSS – with John Barr as valuer – attempted to persuade a ring of firms that had been sharing information with one another for several years, and for which he was acting as secretary, to sell off some of their yards to rationalise the industry and reduce excessive competition! Among the smoke and mirrors, where some of the meeting members knew of the existence of a ring and some did not, the meeting ultimately failed to persuade firms to sell, but it did lead to an important change in NSS strategy. Following a suggestion from Barr to break the stalemate, the NSS agreed to consider the purchase of individual sites or even single berths from a warship firm rather than the entire establishment, if it helped achieve the organisation's goals.[36]

One of the first deals struck under this system was for three yards from Palmer's shipbuilding company (Amble, Hebburn, and Jarrow where the firm was headquartered). Ironically, Palmer's was one of the only firms not present at the meeting when the agreement was reached, and, as the company was in receivership by 1933, what started as a discussion of the purchase of an individual yard ended with the sale of the entire firm.[37] This was soon followed by purchases of berths from Vickers Armstrong (which constituted part of the holdings of the former Armstrong-Whitworth Company), Swan Hunter and Fairfield. Fairfield turned out to be a contentious issue, as Montagu Norman himself felt that it – which needed nearly double its agreed bank overdraft limit to survive – was too important to liquidate. The Bank of England – through its Securities Management Trust (SMT) company – therefore agreed to offer the £200,000 necessary to keep Fairfield in business, but only on the condition that the NSS would buy the company at a later stage if Fairfield could not repay the loan.[38] The final arrangement was for the sale of four of the company's berths instead of the whole firm, but with the personal agreement by Lithgow for the NSS to purchase the rest if "liquidation proved necessary."[39]

By the time Lithgow, Weir and Balfour began advising the CID, the NSS had purchased or agreed to purchase around 800,000 tons of shipbuilding capacity from over twenty firms, comprising some 130 berths at a cost of almost £1.5 million. Although the NSS would continue until the outbreak of war, its work in less than four years to 1934 comprised over sixty percent of the total carried out under the scheme and cost fifty percent more than the value of the original loan by

36 NMM, NSS, meeting of Warship and Liner Builders, 23 February 1933.

37 Hebburn was eventually sold to Vickers Armstrong.

38 BOE, SMT/3/282, Norman to Duncan, 9 June 1933.

39 Slaven, *British Shipbuilding,* 101.

the BIDC. Perhaps more significantly, every one of the NSS's major acquisitions of naval facilities had already taken place or was under discussion by 1934. Yet at no point had attempts to persuade warshipbuilders to sterilise berths at the same rate as cargo manufacturers been particularly successful. The original intention had been to cut around one-third of total shipbuilding capacity – 1.3m tons – from the sector.[40] But by the time the scheme concluded, only thirty-seven of the desired fifty warship berths had been eliminated from the available pool – equivalent to about twenty percent of total capacity – while nearly 180 cargo and merchant berths had been removed against an initial target of 120, equivalent to around forty percent of the available cargo manufacturing capacity at the beginning of 1929.[41] Thus, while the NSS undershot on warship and liner berths, it ended up sterilising over 1.4m tons of capacity in total by 1939.[42]

On the face of events, this should not have been the case. Despite a global economic downturn which hit international trade and personal incomes (which drove the market for merchant and passenger vessels), firms involved in building cargo vessels were also more likely to have a diverse customer base and smaller overheads than those which derived a large proportion of their income from naval contracts.[43] One explanation for this imbalance is the existence and operation of the WSBC. As a mechanism for avoiding cutthroat competition and for raising prices (and profits), the WSBC allowed the warship sector more scope to mitigate the impact of market forces in a way that smaller merchant builders could not. As such, it made them less likely to be enticed by the NSS scheme. Moreover, the long-term nature of disarmament had led the naval manufacturers to create and develop structures to deal with excessive competition some years before the onset of the global economic crisis. While they could also compete for merchant work, the reverse was not true for merchant builders. The WSBC's success in raising the price floor for the relatively few naval orders acted as a disincentive for its members to sell to the NSS, given that typical prices offered under the scheme were just half the pre-war value of a firm.[44] It is, of course, almost impossible to prove why far fewer of the WSBC members could be persuaded to sell from surviving documents – many conversations between

40 NMM, NSS. Minutes, 23 July 1934; and BOE, SMT/2/280, file 142, "Servicing the Loan," 10 November 1930.

41 Compiled from Slaven, *British Shipbuilding,* table 4.9 and NSS, minutes (Slaven notes). Even if one accounts for the latter's smaller average tonnage capacity per berth (around 6000 vs. 9000 tons), the total cargo tonnage sterilised vastly outstrips that of warship (and large liner) work by a ratio of well over 3:1. Incidentally, Slaven calculates the sterilised warship berths at forty-one – I have not included smaller berths, including those at Palmer's yard in Amble (c. 1500 tons each), as although part of a warshipbuilding firm, these were not used to build naval ships.

42 Slaven, *British Shipbuilding,* table 4.9.

43 *Ibid.*, table 4.7.

44 NMM, NSS, minutes, 31 July 1930.

WSBC members appear to have been deliberately destroyed. Thus, the role John Barr played in his simultaneous capacities as valuer to the NSS, secretary of the WSBC and chair of the Shipbuilding Conference remains intriguing, but largely unknown.

Nevertheless, despite the failure by the NSS to reduce naval capacity by as much as it had hoped, the naval sector had been considerably slimmed down compared with a decade earlier. Though some of this had taken place in the 1920s – Vickers and Armstrong-Whitworth merged in 1927, while the Coventry Ordnance works ceased operations in 1925 – it was the NSS which had eliminated most. Beardmore and Palmers were removed from the pool entirely, while Vickers Armstrong, Fairfield and Swan Hunter all reduced the number of berths in use. In all, over 300,000 tons of naval capacity was sterilised by the scheme before the beginning of rearmament (the merchant figure was over 1,100,000).[45] Moreover, for the manufacture of certain components there was a very substantial loss of productive capacity, to which berths and tons alone cannot provide a full picture. For example, the collapse of Beardmore in 1930 removed advanced facilities from the available pool – including some work for naval guns. This was immediately noticed by the Admiralty, whose representative noted the "bottleneck" in gun mounting effectively dictated the pace of all naval construction, as ships could only be brought into service as fast as guns could be made.[46] Similarly, the Beardmore engine works – entirely on the Dalmuir site – is another (if perhaps less dramatic) example of sterilisation that had a greater impact than the reduction in berths alone would suggest.

Lithgow, the Supply Board and the NSS, 1933–1934

The activities of the NSS were thus of great significance to the ongoing planning in the PSOC's subcommittees after Manchuria. For while the NSS was busy from 1929 formulating methods to deal with a perceived overcapacity in the industry during an intense industrial depression, the job of the PSOC and its CID parent was to look as far ahead as possible to assess security threats and take early steps to deal with them. The sterilisation of the Beardmore and Palmer along with parts of the Fairfield yard coincided with the Supply Board's discussions of British naval deficiencies. The problem, however, was that the PSOC was dealing with hypothetical and ever-changing scenarios which were by their nature top secret, and thus could not act publicly to step in and counter the rationalisation schemes to preserve capacity.

In addition to secrecy, there was plainly no money or political will for intervention in industry. Providing orders to keep facilities at Beardmore and

45 Slaven, *British Shipbuilding*, 102.

46 TNA, CAB60.30, 24th meeting of the Supply Board, 11 October 1932, 13.

Palmer in operation would not only have been costly but also could also have contravened the limits set by the 1930 London Naval Treaty and would have sent the wrong signals to Britain's friends (or sparked a retaliation from potential enemies) and the British public, whose pacifist tendencies were reaching a collective peak by 1933.[47] Indeed, a tentative enquiry by the SEF to the Admiralty in December 1930 for an acceleration of naval construction had been rebuffed for these very reasons, with the First Sea Lord concluding that any move "would be overwhelmed by public as well as [parliamentary] criticism to the effect that there were far better ways of spending money than on warship building."[48] While the threats in 1933 were of a quite different and more serious nature, there was still no immediate prospect of a change of heart.

In sum, Lithgow's invitation to sit on the Advisory Panel of Industrialists in the winter of 1933 was most likely closely related to the NSS sterilisation programmes, and perhaps even his role (albeit minor) at Beardmore. Thus, while the PSOC and its subcommittees apparently remained oblivious to the existence of the WSBC price-raising mechanism, they were acutely aware of the possibility of losing firms from the list of suppliers during an industrial depression of this magnitude.[49] Firms that collapsed could, in theory, be restarted relatively quickly if the need arose because their physical assets remained intact, but under the terms of the NSS the assets were quickly stripped and sold off, and the land usually repurposed and prevented from being used for shipbuilding in the future.[50] Without a change of strategy, the crisis surrounding the lack of naval guns and armour-making facilities might even deepen further. In short, although there were many other good reasons for choosing Lithgow as an advisor, his role as Chairman of the NSS meant that, despite not being a naval shipbuilder at the time of his appointment, he was central to one of the PSOC's greatest problems and key to the solution.

It should therefore be of little surprise when it is recalled that the Supply Board's priority for Lithgow, Weir and Balfour when they joined the Advisory Panel was for them to be familiarised with the procurement structure for war supplies so that an accurate picture of British deficiencies could be established. All three men were then supplied with a list of the "key firms" as extant in 1933 as well as the tendering and procurement processes for the War Office, Admiralty and Air Ministry. The content and reception of the February 1934 report was the result of these developments. But it also conditioned what happened within the NSS, which had by early 1934 recently finished negotiating the purchase

47 Kyba, *Covenants without the Sword*, 26.

48 Slaven, *British Shipbuilding*, 90.

49 TNA, CAB4/22, PSO 1109, "PSOC Report: Private Armaments Industry," 31 March 1933.

50 J. Greaves, *Industrial Reorganisation and Government Policy in Interwar Britain* (London, 2005), 223–225.

of the Palmer yards and part of the former Armstrong-Whitworth holdings of Vickers-Armstrong.

Instead of liquidating the assets, as had been a central tenet of NSS policy since the scheme was first discussed in the 1920s, Armstrong's large Tyneside Walker yard, disused but formerly engaged in naval construction, was purchased for £125,000 in July 1934 and put on a "care and maintenance" basis. This meant that rather than being permanently sterilised, it was mothballed with the possibility of re-starting the facility at six-months' notice. The official reason given by Lithgow was that the Walker yard could be used as a "threat" to those who were slow to rationalise, as "restoring a yard such as this could be used to make other builders ... more likely to reduce their facilities."[51] Likewise, the Hebburn yard of Palmers, also a former naval facility, was bought for £100,000 and not sterilised but sold on ten months later to Vickers for a sum of £85,000.[52] Put another way, Vickers-Armstrong had sold one yard that the firm did not want to the NSS and bought one from it that it did, and made a profit of £40,000 in the bargain, while the NSS lost £15,000.[53]

It could be argued that mothballing was an effective way to encourage other naval manufacturers to increase the pace of their own internal rationalisation schemes. There are problems, though, with such an analysis. First, sterilisation of naval berths and facilities lagged that of their merchant counterparts; therefore, retaining a large facility in 1934 when there was still (a perceived) substantial overcapacity in naval construction makes little sense. Moreover, it is hard to see on what basis mothballing would have been effective at enticing rivals to reduce capacity. The NSS would surely have undermined its own argument by bringing another facility back into operation while simultaneously claiming that more berths and yards needed to be closed. Secondly, a complete reversal of a central policy on the back of nearly four years of sterilisation is extremely difficult to explain with the rationale given. Before 1934, there is no evidence of any belief in the wisdom of mothballing yards (at the NSS's expense) or selling yards at a loss to be kept in use – this policy was without precedent and came at Lithgow's suggestion out of the proverbial blue. The notable precedent that *does* exist, however, is the case of the shell-filling factory in Hereford, which Dowding had suggested be mothballed in a communication with Lithgow three months earlier, at the beginning of 1934. Finally, although not known at the time, the Walker and Palmer acquisitions represented the last serious naval purchases by the NSS – some two years before rearmament orders began flowing. In sum, if Lithgow was indeed serious in his attempts to increase the pace of capacity reduction in May 1934, his plan backfired spectacularly.

51 NMM, NSS, minutes, 23 July 1934.

52 NMM, NSS, minutes, 27 March 1935.

53 NMM, NSS, minutes, 12 October 1933, 30 May 1934 and 27 March 1935.

The Walker and Hebburn mothballing cases require an alternative explanation. There is no doubt that Lithgow's thinking changed between the end of 1933 and the summer of 1934, precipitating a fundamental shift in NSS policy. The question is, what caused this shift? When Lithgow's involvement with the NSS is set against the information he gained from the Supply Board, it appears far more convincing that the reasons for mothballing the Walker yard had less to do with other builders being slow to rationalise than it did with the Supply Board's concerns over deficiencies and available capacity. After all, the complete and unexpected change of course only occurred after the Supply Board's fears – and what would be needed to meet the hypothetical defence requirements of the British state – were laid bare to Lithgow. Of course, he could not articulate these publicly, even to his colleagues and fellow board members on the NSS. In short, it is very likely that the NSS, formed as a response to overcapacity in industry and funded by a large Bank of England loan, through the actions of one influential man, deliberately slowed the pace of naval sterilisation. This only happened after the deficiencies in defence were articulated by senior Army, Navy and Air Force members on the Supply Board. While there is nothing to suggest that he was instructed or pressured into this course of action, once the gravity of the situation was made clear, he adjusted accordingly. This not only highlights the role Sir James Lithgow could play in industrial reorganisation but also underlines the effect the planners in the PSOC and its Supply Board could have, albeit indirectly, upon industry through interactions with a small coterie of key industrialists.

Conclusions:
A Changing Political Economy

Lithgow knew much more than the rest of the NSS board (and the WSBC) about the nature of defence deficiencies, but he could not, and did not, share that information. The same is broadly true for Balfour and Weir. This information was nowhere near complete: the CID and PSOC knew a great deal more about the basis upon which these deficiencies were being calculated but were not prepared to share that information with the Advisory Panel in early 1934. Then there was the WSBC, which in 1934 did not contain any of the panel's members and was simultaneously quietly operating its own price-fixing scheme of which most of the directors of the NSS – and the Admiralty, PSOC and Advisory Panel – knew nothing. At the root of this behaviour lay the potentially serious conflicts of interest that no party could risk making public: WSBC members had a substantial presence on the NSS board, which was trying to restructure the market, while the Advisory Panel had access to officially secret information about defence deficiencies and the PSOC's investigations into the businesses of their rivals.

The changes that occurred in these years – both in supply organisation and in industry – shaped the rearmament that followed and preceded the Defence Requirements Committee (DRC) and post-1934 era, which has been characterised as where defence policy began to shift.[54] Compared with 1931, however, there had already been a complete change in war planning and the attitude of the CID. Criticisms of the slow pace of change are still valid, but one should not ignore the major developments that did take place. While the relationship between state and industry extended only to a small handful of men in 1934, the fact they were brought into the confidence of the PSOC, with Cabinet approval, and given top secret information marked a paradigm shift in thinking. Rather than industrialists being outsiders developing responses to crises, they were part of the fold with information far exceeding their peers. Weir, Lithgow and Balfour were now the most important industrial voices on armaments manufacture.

Financially, Britain was still struggling to emerge from the effects of the depression, and politically the government remained reluctant to authorise any increase in armaments manufacture lest it be perceived internationally as aggressors or be subject to a public backlash for abandoning disarmament. This was the climate which existed without much change until 1935.[55] These constraints should not be forgotten, for they explain why the process of industrial involvement evolved as it did. To this end, there was both a huge shift in thinking between 1931 and 1934 and a distinct lack of progress. The financial and political pressures meant that the members of the WSBC and Admiralty, as organisations, still lacked influence over spending in much the same way that they had between 1922 and 1931. The lack of orders was now widely accepted within the CID as having a detrimental effect on Britain's ability to wage war and, most importantly, a secret state-industry framework now existed to investigate the problem. The problem was top-level government, and the next step in beginning to solve the problem could not be taken until there was a sea-change in the political (and financial) climate.

---

54 McKercher, "From Disarmament to Rearmament," 21–22.

55 Shay, *British Rearmament,* 46–47.

1. Lord Weir.
Source: Glasgow University Archive Services (GUAS), Lord Weir Collection, GB248, DC 96.

2. Arthur Balfour (Lord Riverdale). Taken from *The Engineer,* 1936.
Source: GUAS, William Beardmore and Co. Collection, GB248, UGD 100.

3. Sir James Lithgow (left) and King George VI with staff at Beardmore, 1938. *Source*: GUAS, William Beardmore and Co. Collection, GB248, UGD 100.

4. *opposite* One week's typical output of four-inch naval guns, 1920, Parkhead. *Source*: GUAS, William Beardmore and Co. Collection, GB248, UGD 100.

5. View of the turbine and gearing workshop during the 1920s, Dalmuir.
Source: GUAS, William Beardmore and Co. Collection, GB248, UGD 100.

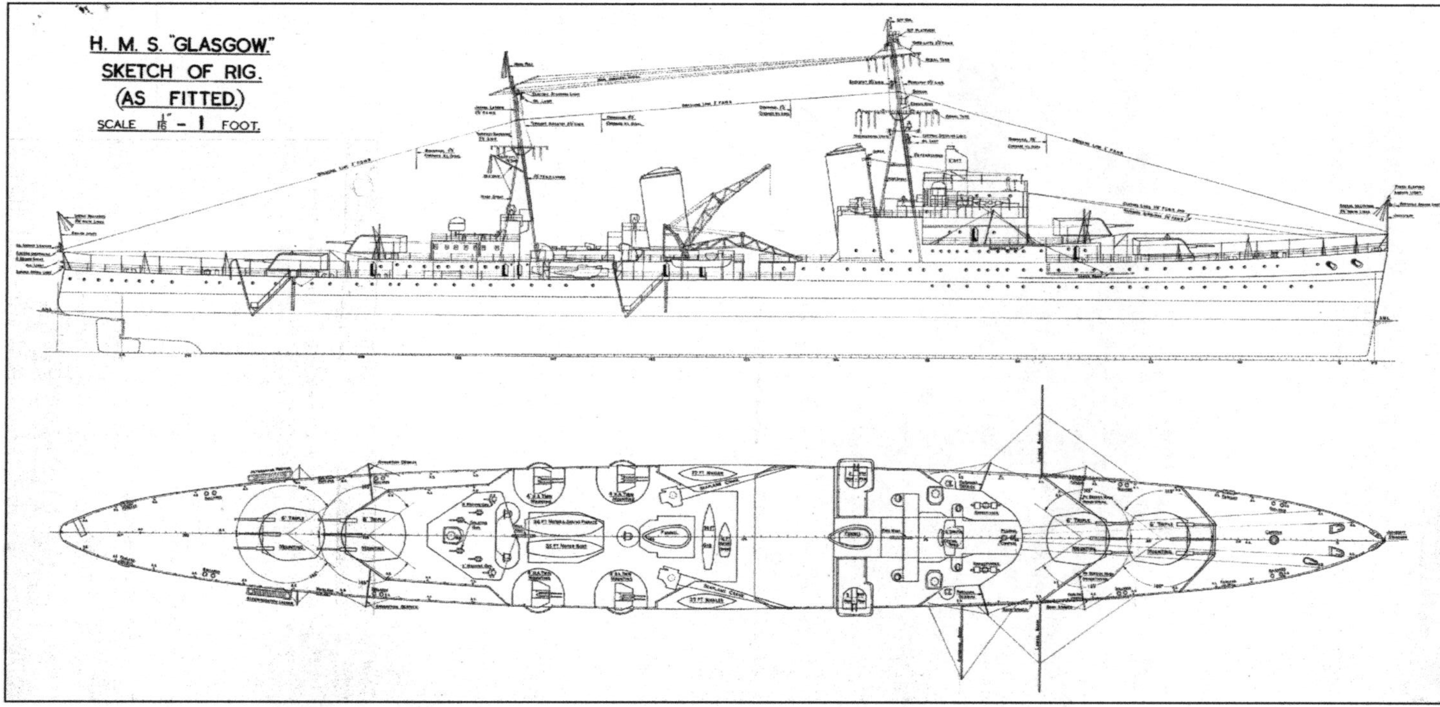

6. Sketch of HMS *Glasgow,* 1936.
Note: *Glasgow* was part of the last class of cruisers to be built to the Washington-era restrictions.
Source: National Records of Scotland and GUAS, Scotts Shipbuilding and Engineering Ltd. Collection, GB248, GD 319.

7. Front cover of Scott's brochure showing the Greenock Yard, 1932.
Source: National Records of Scotland and GUAS. Scotts Shipbuilding and Engineering Ltd. Collection, GB248, GD 319.

8. HMS *Duke of York*, one of the *King George V*-class battleships, during fitting-out at John Brown's Clydebank works, 1940. *Source*: GUAS, John Brown and Co. Collection, GB248, UCS 1.

9. Pneumatic riveters working at Lithgows, c. 1930. *Source*: GUAS, Lithgow's Collection, GB248, GD 320.

10. Engine shop no. 2 bay at Beardmore's Dalmuir Yard during the 1920s. *Source*: GUAS, William Beardmore and Co. Collection, GB248, UGD 100.

Part Four

The Ultimate Potential Enemy and Rearmament Planning, 1934–1936

Chapter Seven

Towards Rearmament, 1934–1935

By the end of 1935, Britain's international security looked far less certain than it did in the spring of 1931 because now there were three credible adversaries: Japan, Germany and Italy. As a direct result of these developments, the tentative work of the CID to explore, if not tackle, industrial and material deficiencies that began after Manchuria took on even greater meaning and urgency. By the time that Mussolini's forces had invaded Abyssinia (modern-day Ethiopia) in October 1935, a two-front war in Europe and Asia requiring a renewed navy and an expanded air force was considered by the CID to be a very real possibility. Moreover, the idea of a major rearmament effort entered, finally, into in public consciousness.[1]

Political Background

Of course, the first indications of a two-front threat during this time came not from Italy but from Nazi Germany, which had emerged as a force under Hitler in early 1933 and withdrew from the Geneva disarmament conference that October. The deterioration of relations with Germany prompted the CID to formally plan based on a twin German-Japanese threat by the beginning of 1934.[2] Cabinet responded to the CID's concerns with the largely Hankey-led initiative known as the Defence Requirements Committee – another CID subcommittee. The decision by Cabinet to grant the DRC approval to examine the "worst problems in British defence" has been characterised by some historians as the "first major step" in war planning since 1918 and the "beginning of rearmament."[3] As to the first of these contentions, it was nothing of the sort. While it may have been an incremental change in top-level foreign policy insofar as it was a clear recognition by Cabinet of the need to improve the condition of British defences, as far as planning is concerned it was only a continuation of the work of the

[1] Kyba, *Covenants without the Sword,* 152–155.

[2] P. Bell, *Chamberlain, Germany and Japan, 1933–1934: Redefining British Strategy in an Era of Imperial Decline* (London, 1996), 16.

[3] Neilson, "Defence Requirements Subcommittee," 651–652.

supply subcommittees. Deficiency hypotheses had been circulated since 1931, so while the recognition of the German threat at the end of 1933 – and the formation of a committee with an explicit remit to discuss deficiencies – has perhaps understandably been taken as a point of departure for the study of rearmament policy, in terms of *industrial* deficiencies and planning the work was already well under way, as evinced by the long process which led to the establishment of the Advisory Panel.

The PSOC and DRC's work here led to the abandonment of disarmament and the adoption of rearmament, announced to the public through two White Papers on defence in March 1935 and March 1936. At Cabinet level, however, the general theme before 1937 was one of indecision, time wasting and missed opportunities, which together led to a failure to act upon the PSOC's planning framework to materially remedy deficiencies. The financial and political constraints were beginning to ease by late 1934, but the continued fear of public outcry or economic collapse underpinned the thinking of senior members of the National Government, hampered the progress of defence planning, and led to frequent divisions over the nature, speed and cost of any future defence programme. In short, there was much good work taking place within the lower echelons of the CID, but it lacked overall direction until well into 1936.

Moreover, while the background to the White Papers is well known political history, it does have an industrial element which is too often overlooked. Yet industry was central to defence policy. Firstly, the industrial problems – initially facilities, and then materials and skilled labour – were always present in defence policymaking after 1934, and crucially were increasingly understood through the advice given by and experience of the CID's civilian industrialist members. These men, particularly Lord Weir, came to play important roles in articulating defence plans. Secondly, there is the issue of profits. It was towards the end of 1935 that rearmament "went public," and while the secret work of the PSOC and Supply Board continued, the idea that Britain was going to spend large sums on re-equipping her fleet and building an air force was widely understood as imminent. Therefore, the activity – and thus value – of established armament makers began to increase once orders started being placed in 1936, and the public, which retained pockets of firm pacifism, openly questioned whether war should lead to private gain.[4] The subsequent Royal Commission on the Private Manufacture of Armaments, which published its final report in October 1936, was the National Government's public acceptance of such feelings.

The worsening geopolitical situation finally compelled the National Government to rearm and fundamentally transformed the outlook for armament manufacturers. These chapters highlight the naval dimension of the discussions and argue that despite the Royal Navy retaining a central position in defence planning, naval rearmament was still a slow and tentative process. Indeed, for

4 Anderson, "British Rearmament," 15.

private manufacturers it only began to gather pace in latter half of 1936 – at which point the WSBC's members began to reap the fruits of their 1920s collaboration in a huge way. Most importantly, however, rearmament considerably raised the standing and influence of Balfour, Lithgow, and particularly Weir, and all three benefitted materially from their work in the CID. Of course, many other manufacturers not connected with the CID and not privy to the same level of "inside" information also received increased orders. The crucial difference with the members of the Advisory Panel, most notably Lithgow, was that between 1934 and 1936 he managed to transform his business portfolio from a merchant shipbuilder to an armaments magnate in a quite astonishing and unique way.

Industrial Background to the White Paper: The DRC and Abandoning Disarmament

Much like the Ten Year Rule recommendations in the wake of Manchuria, it was a CoS strategic review, this time in November 1933, which highlighted a growing threat – now Germany – and quickly passed through the CID and the Cabinet. The final Cabinet recommendation was for the formation of a new subcommittee, the DRC, which would be comprised of the CoS plus Hankey (the Cabinet and CID's chief civil servant) along with Vansittart and Fisher, the chief civil servants in the Foreign Office and Treasury, respectively (see Appendix 3).[5] Although generally contemporaneous with the Advisory Panel, the work of the two did not have direct relevance to each other until the summer of 1934. Supply deficiencies were a separate question stemming from discussions in 1932, and the work of approaching industry had its own, longer roots which predated developments in Germany. For these reasons, the Supply Board had its own separate history until the publication of the first DRC report in mid-1934, but thereafter it was strongly guided by the recommendations laid out by the DRC's committee members.

The DRC is often credited with identifying Germany, rather than Japan, as Britain's "ultimate potential enemy."[6] But the October report of the CoS, which recommended establishing the DRC in the wake of Hitler withdrawing from the Disarmament Conference, had in order of importance, "the defence of the Far East, European commitments, and defence of India from Soviet aggression" as its top priorities. Moreover, it found that naval expenditure, particularly on cruisers, should be increased to meet Britain's commitments.[7] Of course the Treasury and the Chancellor (from 1931), Neville Chamberlain, did not see things this way

[5] Neilson, "Defence Requirements Subcommittee," 662.

[6] *Ibid.*, 652; and W. Wark, *The Ultimate Enemy: British Intelligence and Nazi Germany* (Ithaca, NY, 1985), 23.

[7] Neilson, "Defence Requirements Subcommittee," 662; and Wark, *The Ultimate Enemy*, 23.

and hoped that the DRC would be sufficiently well represented by Treasury and Foreign Office officials so that the views of the CoS – which had a "tendency to add up the requirements of all three departments rather than establish priorities" – could be tempered.[8]

While the deliberations of the DRC in formulating defence hypotheses were certainly relevant to private industry (insofar as the composition of future equipment orders would be shaped by the answers given), covering every detail of the committee's discussions over the next three months would take us away from our central narrative. Still, there are some aspects worth highlighting. First, in the weeks prior to the DRC's first meeting, the Treasury, through Warren Fisher, had insisted that the DRC should be clear on which nation or combination of nations posed the greatest danger *before* any scheme to prepare British defences was drawn up.[9] This was an important point, for it linked industrial and military preparedness for war to the political and strategic decisions about the nature of the threats faced. It was *not* assumed that Britain would consider threats and spend money on building up the foundations of rearmament concurrently; industrial preparation was only to follow once the National Government could agree upon the direction of policy. In practice, this caused problems and delays until at least 1936, for there remained myriad different views within the DRC and Cabinet about Britain's "ultimate potential enemy."[10]

Secondly, when the DRC first met in late 1933, the Foreign Office representative, Sir Robert Vansittart, had already accepted Treasury's views on the limits to defence expenditure and seemed concerned especially with the weight of domestic feeling against armament orders. He argued that:

> I should prefer to guard against [all the risks], but we obviously cannot do so. It would cost far too much money and far too many votes apart from numerous other considerations. If we cannot cover the whole ground, first things come first, and we must begin a day's march nearer home.[11]

Financial and domestic political considerations were guiding the Foreign Office official's thought, in much the same way as they had guided most government decisions throughout the 1920s to reign in armament spending and continue pursuing disarmament. Interestingly, it appears from Vansittart's words that preparing against Germany was as much about practicality as about ideology. He had argued previously that Britain could do nothing about Japan if she desired to attack, so to him the logical progression was to allocate most of the resources

8 See *ibid.*, 682 for more on this; see also Peden, *British Rearmament and the Treasury,* 28.

9 Peden, *British Rearmament and the Treasury,* 109.

10 Wark, *The Ultimate Enemy*, 23.

11 TNA, CAB16/109, 2nd DRC meeting, 27 November 1933.

to areas where they would make the most difference, not where the threat was currently largest or the deficiencies greatest. Secondly, the important underlying assumption was that this was a discussion of priorities rather than deficiencies, and this became the guiding factor in early DRC deliberations. At the first meeting, both Hankey and Fisher referred to the question of priorities explicitly, while Vansittart reiterated his view that Europe must come first – although he admitted that this view was not widely held among other Foreign Office staff.[12]

Essentially, the rest of the discussions followed a pattern not dissimilar to Admiralty-Treasury disputes in the 1920s. Chatfield continued to highlight the danger in the Far East, the relative weakness of the Royal Navy, and the reliance on sea power for supply routes in the event of war. On this he was broadly supported by Hankey. Vansittart and Fisher believed Japan would not attack Britain unless out of opportunism, in other words when Britain was fighting another power. To Vansittart and Fisher, this meant that Europe was the most likely place to be "otherwise engaged," and it followed that Britain should focus on Germany. To Hankey and Chatfield, the same line of reasoning meant that *if* Britain became embroiled in a European conflict, Japan would sense weakness in the Far East and be more likely to attack; as such, the two were related questions and should be considered as equal priorities.[13]

What followed could as well have been from the "cruiser crisis" of 1925 or the immediate aftermath of Manchuria. Hankey and Chatfield felt the answer was to increase naval expenditure to deter Japan, while the Treasury view was that no amount of expenditure in Singapore and Hong Kong would make British interests any more secure. The other two members of the committee, representing the Air Force and War Office, backed Chatfield. Their view was that Germany remained some years from becoming being a realistic military threat to Britain, so Japan should be the focus of current attention.[14] Despite being four against two, the DRC had, in early 1934, split along clear lines – the CID secretary and three Chiefs of Staff on one side, and the chief civil servants of two major government departments on the other.

The DRC met another six times in the first two months of 1934, but the gap between Fisher, Vansittart and the CoS and Hankey did not close. It is for this reason that the final report, authored by Hankey, had elements which helped placate the demands of both sides, but was thus merely a compromise that was far from achieving the "clarity" the Treasury desired. It recommended that Britain seek a "policy of accommodation with Japan" and that Germany was the "ultimate potential enemy."[15] These were both weak characterisations which

---

12 TNA, CAB16/109, 1st DRC meeting, 14 November 1933.

13 TNA, CAB16/109, 4th DRC meeting, 18 January 1934.

14 For a discussion of the various points made in the debate, see Neilson, "Defence Requirements Subcommittee," 665–673.

15 *Ibid.*, 672.

stopped short of recommending that full attention should be given immediately to Germany, but nevertheless were enough to get Vansittart and Fisher to be signatories. It did broadly follow the lines set out by the CoS and particularly by Chatfield; in terms of deficiencies, Singapore and the Far East were priorities, and the Royal Navy was to receive an extra £21m to meet their immediate deficiencies, and would need to spend more than £14m a year for the five years until 1939 to replace and update older ships and provide them with aircraft support – in other words another £94m of funding before 1939.[16]

Chamberlain and Naval Defence Spending

The DRC's report, like the PSOC's request to approach industry, required ministerial approval. In the case of the DRC, this was set for May 1934 in the Ministerial Committee on Disarmament.[17] Unlike the PSOC's request, however, this issue prompted a heated and prolonged debate that in some ways was a repetition of the discussions in the DRC. The Treasury (this time represented by the Chancellor, Chamberlain) rejected the findings of the report and argued for a reordering of the DRC's priorities with a focus on a massive aerial deterrent and no increased naval work. Chamberlain based his views on the public fear of Germany and a perceived public preference for increases in welfare spending over increases in defence spending. Moreover, he insisted that a balanced budget was a priority.[18] Chamberlain found some sympathy from MacDonald and Baldwin on cost reductions and support on ruling out borrowing to fund any new defence programmes, but neither wanted any further cuts in naval expenditure.[19] Others were divided on principle: accepting a large increase in defence expenditure would derail the upcoming naval limitation talks[20] (and, it could be argued, perhaps even the disarmament process in Geneva) permanently and wanted Britain to explore all limitation avenues before considering rearmament.

In short, there were still wide differences in opinion between the Treasury and Admiralty that could not be solved easily. It was August 1934 before agreement was reached. At the final meeting the Admiralty received some last-minute support from several ministers and civil servants, including MacDonald, Baldwin and Anthony Eden (the latter at this point the Under-Secretary for Foreign Affairs). All three believed that friendly relations with Japan were likely to antagonise America, and that to remain a presence in the Far East, Britain had

16 Gibbs, *Grand Strategy,* 123. This was divided into £21m for rectifying deficiencies, £67m for new naval construction, and almost £6m for the expansion of the Fleet Air Arm.

17 *Ibid.,* 99.

18 Peden, *British Rearmament and the Treasury,* 68–69.

19 McKercher, "From Disarmament to Rearmament," 39.

20 Gibbs, *Grand Strategy,* 126.

to rely on the Royal Navy.[21] The British fleet was demonstrably unable to meet this commitment: if any substantial fleet was sent to Japan, the remaining force at home was no longer large enough to protect trade routes and engage a rival power in European waters.[22]

Inexplicably, Chamberlain was allowed to draft the final agreement along broadly the same lines he had been proposing all along without much interference from his opponents (including Hankey who, in his capacity as Cabinet Secretary, drafted the original DRC report). Given the heated disagreements between Treasury and Admiralty and the fundamental disagreements between ministers, how this could happen, despite Chamberlain's power and influence, is still shrouded in mystery.[23] Thus, citing "overriding financial considerations," Chamberlain succeeded in vastly reducing the increase in defence expenditure from £71m to £50m. This included a cut in the Admiralty's share of the increase by £8m (to £13m), halving the Army's share to £20m and doubling the increase the RAF would receive to match the Army's level. The 1934–1939 naval replacement programme was also altered, no longer to be considered as a five-year programme but now to be dealt with on an annual basis – a notional cut of more than £60m from the estimated £67m ship replacement scheme.[24]

Peter Bell has called this a "major change of emphasis" in favour of the RAF, while Keith Neilson has taken this episode as an example of Treasury dominance over the Admiralty budget, contrasting this with the 1920s when the "Admiralty had managed to fend off [the Treasury] using a combination of political strength and need to respond to international events."[25] It difficult to support either view fully. Firstly, too much has been made over the revision to the five-year replacement plan. In the 1920s, the Treasury used the relatively settled international situation as a reason to successfully restrain naval expenditure – between Washington and the onset of the financial crisis eight years later, Admiralty expenditure remained remarkably stable: in 1922, it was £81m, and thereafter it stayed close to £55m, with fluctuations of no more than five percent either way (excepting the one-off increase negotiated for cruisers in 1926), until 1932. Moreover, the replacement plan in the initial DRC report was supposed to be an estimate based on previous needs, and both its members and Chamberlain noted that these could be amended if the situation changed in the future, particularly

21 Neilson, "Defence Requirements Subcommittee," 675–676.

22 McKercher, "From Disarmament to Rearmament," 37–39.

23 Neilson, "Defence Requirements Subcommittee," 677, argues that the "need to get a final agreement was politically pressing" and that the Foreign Secretary, Sir John Simon, probably agreed because he believed more money could be added subsequently.

24 Gibbs, *Grand Strategy,* 124. Bell wrongly asserts the cut was from £67m to £55m. It was actually to be cut to £5.5m.

25 Bell, *Chamberlain, Germany and Japan,* 124; and Neilson, "Defence Requirements Subcommittee," 678.

with regards to a German threat.[26] It was therefore a moot point – most of the replacement money was earmarked for 1937–1939 – and as the geopolitical situation had changed markedly by then, it meant that the rest of the funds and much more besides was in the event allocated to the navy.

In addition, Neilson has pointed to "industrial constraints, not finance" as the determinant of the speed of naval construction, implying that because capacity was limited, a major naval rearmament programme could not have been undertaken in 1934, even had Chamberlain wanted.[27] The point about industrial constraints was generally true – indeed, it was the focus of Supply Board investigations – but it does not hold here. The Supply Board, unlike the DRC, was looking at capacities and facilities assuming a major war where production would need to be maximised. This was not the scenario in 1934. The private armaments sector had been thinned down since Washington, but it still comprised several large (and largely unused) naval shipyards. Even following NSS closures, Slaven calculates that around 100 berths suitable for warship or liner construction, with a total capacity of three-quarters of a million tons, survived into 1935.[28] By way of comparison, total private warship construction was barely 80,000 tons across the years 1933–1935.[29]

This highlights an important point: even the full £21m on top of the 1934 budget was well below expenditure levels in the previous war and could have been absorbed by existing private industrial capacity in peacetime (the *King George V* programme for a class of five ships was nearly £40m in 1937).[30] Given that the DRC was looking at least five years into the future for any expenditure programmes to bear fruit, it was not necessarily total spending but how the money was to be spent that was the issue. In the case of the major supply bottlenecks, such as gun-mounting and armour, Chamberlain's proposals effectively put off immediate expenditure and waited for circumstances to change. As later chapters in this book argue, extra spending on deficiencies in 1934 could have paid dividends during later rearmament, and the constraints the PSOC and its subcommittees identified could have been overcome and then expanded in the five-year period.

It should also be remembered that Chamberlain's actions are evidence of a slight improvement in the Admiralty's lot compared with the 1920s. Despite rejecting the notion of the Japanese threat, having support for financial prudence and being left in charge of drafting the final agreement, Chamberlain still agreed to a £13m increase in expenditure to remedy naval deficiencies. In other words, the Manchurian crisis had shifted the terms of reference of the debate

---

26 Gibbs, *Grand Strategy,* 124.

27 Neilson, "Defence Requirements Subcommittee," 680. Neilson does, however, note that there were no such restrictions on the "mundane" matters of manning and oil supply, and more could have been achieved here.

28 Slaven, *British Shipbuilding,* 102.

29 Johnman and Murphy, *British Shipbuilding and the State,* table 5.

30 Edgerton, *Britain's War Machine,* 212.

and prompted the National Government to think that conflict of some sort was possible well within the "ten years" that had guided thinking until that point. The discussion in 1934, protracted as it was, did not establish Germany unequivocally as the main threat, but it did work under the assumption that defensive deficiencies existed and that it was now a question of how much extra to allocate to the service departments and how that figure was going to be divided. To focus on Chamberlain cutting £20m from the total in 1934 (and doubling the amount of the RAF's increase) and looking at the headline figures alone misses important nuances. The text of the final agreement reaffirms concerns about Far Eastern defence and the need to show naval strength to deter Japan.[31] Put simply, despite the problem of industrial bottlenecks, the argument to attach a low priority to naval spending could not be won so easily after Manchuria.

This can only partly be explained by the political fall-out of reversing the decision of a Cabinet-appointed committee. The role the PSOC and Supply Board played in establishing the extent of industrial and material deficiencies has almost always been ignored, but it should be remembered that most of those who spoke up favour of increased naval expenditure, including Baldwin, MacDonald and Hankey, had all viewed and been conditioned by the PSOC's warnings of overreliance on a small group of struggling firms and the condition of equipment in Britain and around the Empire. In 1932 and 1933, evidence had been collected to show the seriousness of the industrial problem. Moreover, Britain's industrial situation had been starkly contrasted with that of Japan, Germany and Italy long before the DRC had even been constituted, let alone had ministers met to discuss its report.[32] It was, however, only a compromise. What Britain needed for a future war was investment in industrial capacity, facilities and skills. In this episode, the Treasury had taken the sting from the DRC and succeeded in revising downwards the proposals. It was a small but significant step in the right direction and perhaps a better settlement than otherwise might have been for the Admiralty, but progress was still slow.

Only so much could be achieved by planning and organisation; the rest could only be solved by investment or extra orders that utilised more capacity. In this regard, the service which retained the largest industrial base in peacetime was the Admiralty. Factories could be turned over to shell production; rifles and air frames could, after some start-up time, be mass produced. Warships could not. Therefore, a refusal to grant the navy a significant increase would have reversed almost every supply recommendation the CID or ministers had accepted since February 1932. In short, the PSOC was having some effect in educating ministers on supply issues, but it faced barriers to making the progress it believed necessary.[33]

---

31 Gibbs, *Grand Strategy,* 126.

32 TNA, CAB60/13, PSO 361, "Report by Subcommittee on Industrial Intelligence in Foreign Countries," 7.

33 See Table 7.1. This gap was not closed until 1939. Indeed, despite conventional

Regardless of the reasons for arriving at the cost of the deficiency programme, it should be borne in mind that the DRC's report was not publicly announced as rearmament in July 1934, and discussions took place in secret. In the days before the agreement was finalised, Baldwin pointedly refused to elaborate in Parliament on the deliberations taking place in the Cabinet on defence and reiterated his and the National Government's commitment to disarmament.[34] Once the sums were finalised, the announcement was similarly low-key – Baldwin mentioned only the air programme and omitted any reference to naval or army spending.[35] This was the key conundrum for Baldwin: how to plan war while pursuing peace. The middle ground he took was understandable considering the circumstances, but it was not ideal for defence preparation. This middle ground would be a recurring theme for two more years to come.

Finally, it should not be assumed that this extra £50m was granted immediately to the defence departments and promptly spent to remedy deficiencies. The deficiency programme was explicitly supposed to address the problems across the "five years from 1934."[36] This gradual plan was adhered to – there were no significant spending increases in 1934 or 1935 (see Table 7.1 below). Aside from a limited programme of small, light cruisers, most of the initial DRC money allocated to the Admiralty was only used up in 1936 and 1937 when the geopolitical situation had deteriorated further and new, more expensive rearmament programmes had very publicly replaced the tentative measures set out in the summer of 1934. Indeed, after being cut from over £80m to £56m in the aftermath of Washington, naval expenditure did not go past the £60m threshold again until the year ending 1936. By way of comparison, in 1922 it was £81m (and in 1920 was £157m, although this included demobilisation costs). Spending only got close to pre-Washington levels from 1937. Moreover, expenditure on the Admiralty was *not* the same as the sum spent on new construction. Between 1934 and 1936, for example, around £35m of the total figure was reserved for overheads – costs and wages – leaving only £20–30m of the total available for spending on equipment.[37] Thus, despite the step forward that July 1934 represented for the fortunes of the Royal Navy, this was still very much a period of planning without spending. It may have been the end for disarmament, but it was not yet the industrial beginning of rearmament.

---

explanations of rearmament stressing the prominent role of the RAF, the total amount spent on the Admiralty between 1936 and 1938 was £247.8m, while the RAF received £159.3m. Shay, *British Rearmament in the Thirties*, 297.

34 Hansard, HC Deb, 19 July 1934.

35 Hansard, HC Deb, 30 July 1934.

36 Gibbs, *Grand Strategy*, 96.

37 Johnman and Murphy, *British Shipbuilding and the State*, 57; and TNA, CAB102/4, "Estimated Annual Expenditure on Rearmament, 1934–1939," 3 July 1940.

Table 7.1:
Naval and RAF Defence Expenditure, 1932–1939

Year	Naval Expenditure (£m)	Naval % total defence expenditure	RAF (£m)
1931/2	51.0	47.4%	17.9
1932/3	50.1	48.5%	17.1
1933/4	53.4	49.6%	16.7
1934/5	56.6	49.2%	17.6
1935/6	64.9	47.3%	27.5
1936/7	81.0	43.6%	50.0
1937/8	101.9	39.8%	81.8
1938/9	132.4	33.3%	143.5

Source: R.P. Shay, *British Rearmament in the Thirties: Politics and Profits*, 297.
Note: These figures were taken from Shay's work while earlier figures in this book use B.R Mitchell. This is because Shay's figures are the most accurate available for the 1930s (unlike B.R. Mitchell, he includes expenditure from the Defence Loan), but Shay does not provide figures for the 1920s. It is important to note that Shay counts from the beginning of each financial year – for example 1931 in Shay is 1.4.1931–31.3.1932 – while Mitchell is somewhat unusual and counts from the end. For consistency I have moved Mitchell's years into synchronisation with Shay's, meaning each year is denoted as the year beginning 1 April.

PSOC and Supply Board Developments, July 1934–March 1935

After the first report of the Advisory Panel, comparatively little was achieved or even discussed in the PSOC or Supply Board in the six months that followed. Instead, selecting the "400 or so" major firms that had been suggested as a sensible maximum was the order of the day.[38] It should be remembered that this period coincided more or less exactly with the months in which Hankey's finished DRC report was under review by ministers (and the subsequent parliamentary recess in August and September), so with the CID's underlying defence hypotheses likely open to revision at short notice between April and July, it was perhaps understandable that the focus was elsewhere.

The main development of note before July was the formal acceptance by the Supply Board of the recommendations in the Weir, Lithgow and Balfour

[38] TNA, CAB60/14, PSO 415, "Memorandum by Lord Weir, Sir James Lithgow and Sir Arthur Balfour," 20 February 1934.

report at the end of April.[39] The Supply Board stated that it "concurred wholeheartedly" with the plans to create a shadow armaments scheme and "cordially endorsed" the plans of investigating a smaller list of major firms and placing the "educational orders" with them.[40] Dowding believed the panel's proposals would be "of the greatest value in the field of supply" and that steps should be taken along these lines "as far as is practicable."[41] The Board reaffirmed the tenet of only approaching a very small group of key industry experts as and when required and reiterated that in terms of solving manufacturing problems they should "avoid any suggestion of dictating precise methods' to industrialists and to give them 'as much latitude as possible."[42] While this is a quite remarkable degree of deference to the wisdom of civilians with, after all, only a limited degree of CID knowledge, it is perhaps because the Advisory Panel had not proposed or found anything radically different to what the PSOC had already believed that the Supply Board was willing to give such unquestioning support to its suggestions. Nevertheless, the "latitude" which they proposed to give is noteworthy. As later chapters show, Weir's latitude – and influence – only continued to grow.

There is a related example from 1935 which perhaps helps to explain why Weir, Lithgow and Balfour were so valued and remained part of the CID machinery for the duration of the 1930s. At Weir's suggestion, the PSOC approached Sir Glynn West, Sir Keith Price and Sir William Alexander for consultation on the expansion of shell production. All three had been employed in this capacity in the last war, with West serving as Director General of Shell Manufacture under Weir in the Ministry of Munitions. Like the Advisory Panel, this trio was invited to prepare a memorandum, which was circulated to the PSOC and Supply Board. The crux of their suggestions was a return to the peacetime construction of shell factories and the retention of a nucleus of staff. The Supply Board unanimously found this proposal "quite impracticable" and noted that many of the views expressed were simply incorrect. They were so unimpressed that Dowding remarked that there appeared to be "some disadvantage in obtaining advice from individuals whose experience was confined to the last war or prior to 1914" as they were "not fully *au fait* with existing conditions." The trio was compared unfavourably with Weir, Lithgow and Balfour, who were described as providing "very valuable assistance." The conclusion was merely that the three men "should be thanked for their work" and not be consulted again.[43] This illustrates both how rare the panel's knowledge was – being both "*au fait* with existing conditions" and having

---

39 TNA, SUPP3/18, PSO 421, "Note by Chairman of the Supply Board on the Memorandum by Lord Weir," 25 April 1934.

40 *Ibid.*, 2–3.

41 *Ibid.*, 3.

42 *Ibid.*, 6.

43 TNA, CAB60/31, PSO(SB), 38th meeting of the Supply Board, 4 April 1935; and PSO 467, memorandum by Sir Glynn West, Sir William Alexander and Sir Keith Price, undated.

wartime experience – as well as how rare industrial involvement in the CID's subcommittees remained, even in 1935. It would be well into rearmament itself, when the political and industrial dynamic had changed dramatically, before anything remotely similar was reached by any other group of business experts, and even then, none matched the Advisory Panel in terms of breadth or depth of involvement.

Returning to the situation in the late summer of 1934, once Chamberlain's expenditure agreements had been finalised it was clear that the CID's new (and in some ways, more definite) defence hypotheses had rendered a significant part of the Supply Board's earlier calculations and investigations obsolete. Taken in tandem with the Advisory Panel report, this had a clear impact on the relationships between the individual Supply Committees, as the main task in September for the coordinating Supply Board was to sort out a fresh dispute between SC I (Munitions) and SC III (Shipbuilding). This was over what SC I claimed were "major firms selected for investigation in accordance with Lord Weir's recommendations." The dispute occurred after SC I had prepared a list of several hundred firms that it felt it would need to utilise and submitted it to the Supply Board just two days before it was next due to meet. Many of these firms had been provisionally allocated to or, in some cases, even investigated by other Supply Committees.[44] Dowding, the Supply Board Chairman, called this move by SC I "disconcerting" and believed it should only have "laid claim to those firms which other Supply Committees had not been allocated with a view to the supply of armaments." Witham (SC I and Assistant Director of Factories, War Office) then stated bluntly that a reallocation of firms to his committee "would be necessary, and that the sooner this was done the easier it would be for all concerned."[45]

What then followed was a rare spat between the board's members: Vice Admiral Henderson said the proposal was "impossible to agree to" and demanded the request from SC I be withdrawn, while Witham and Sir Reginald Townsend (Director of Ordnance) drew entirely different interpretations of what Weir meant by his "400 firms." Witham clearly believed that because the industrialists were first consulted over the problems facing SC I in 1933, Weir's recommendations applied explicitly to his committee. Townsend, who also represented the War Office, attempted diplomacy and considered "that it was not Weir's intention that 400 firms be allocated to [SCI] in particular, but rather that they should be devoted to war production, including the requirements of other Supply Committees."[46] Fearing that all previous investigations until that point would be scrapped and the process started afresh, Bovenschen (also from the War Office)

---

44 TNA, CAB60/41, PSO(SB)438, "List Prepared by Supply Committee No. I," undated.

45 35th meeting of the Supply Board.

46 *Ibid.* Witham and Townsend were both from the War Office; Townsend was effectively Witham's direct superior.

argued against Witham's proposals and at the very least that some caveat need be inserted to allow first claim on "essential products." The meeting ended with Dowding and the others siding together and instructing Witham in no uncertain terms to discuss the issue with the heads of the other Supply Committees first and to "determine what capacity could be made available to him, and proceed on that basis."[47] By the standards of other meetings, this was a frank telling-off from Dowding, but it does highlight the role of the industrialists – and the limitations of the Supply Board system – rather well.

Firstly, it is apparent that the Advisory Panel – while usually referred to as just "Weir" – were genuinely treated as experts whose advice was trusted and taken very seriously. At no point was the validity of their conclusions challenged; instead the disagreement was over the interpretation of how the advice of the panel could be applied to the individual Supply Committees. Secondly, the Supply Board in 1934 appeared to be just as concerned with avoiding the various SCs cutting across the bows of one another as it was with allocating capacity in the best way to the relevant bodies. In other words, it sought fairness through enforcing the first claims on capacity (even when it meant members from the same service department arguing against each other) rather than acting as an executive body which decided what each SC was to receive. Thirdly, while the Supply Board was undoubtedly a useful forum for discussing competing claims on industrial capacity, its conclusions were constantly open to revision. The Vickers investigation in 1932 had resulted in a fragile, non-committal agreement among services, but such agreements were prone to shattering when defence hypotheses changed.

By the autumn of 1934 there finally were some notable developments in supply organisation. Building upon the foundations put in place by the Advisory Panel's memorandum, the Board had a sufficient outline of requirements to allow it to focus on more specific topics. When the industrialists were first approached in the winter of 1933, the main issue was the condition of armament industries, particularly the problems facing SC I (Munitions).[48] At the centre of this question were the large private armament manufacturers and the need to produce field and naval artillery shells. Guided by the experience of the Great War, the industrialists had initially spent a significant portion of their time thinking about how the specialist armament makers could take the lead in turning other firms into mass producers of relatively simple items in the event of war (under the headings "elasticity in supply arrangements," "decline in the armaments industry" and "methods of expansion") to avoid supply bottlenecks. For this reason, the decline of firms like Beardmore and the issue of shell manufacture were closely related, as both were considered part of the general problem of the industrial knowledge

[47] *Ibid.*

[48] See TNA, CAB60/40, PSO(SB) 376, "Memorandum by Master General of the Ordnance, War Office," 1 October 1934.

base.[49] By November 1934, the focus shifted from the industrial facilities to the related problem of skilled labour, which had hitherto been only a footnote in the discussions.

It was only at this relatively late point that the PSOC began a series of more definite and detailed calculations of skilled labour requirements, though this would recur throughout the rest of the decade.[50] When the question had first been brought up in 1929, the Supply Board was instructed by the CID to "confine their attention to the requirements of raw materials, with a view to simplifying the problem as far as possible."[51] Thus, although the question had been considered periodically since then, it was only at the end of 1934 that the issue gained sufficient traction to be properly investigated. As *Nelson* and *Rodney* had shown, it was difficult even in leaner times to find skilled workers. More importantly, while the work on these two ships had helped train new staff, by 1935 these ships had been complete for almost a decade, and in the absence of other orders in the interim, much of the benefits of the training had been lost.[52]

The shift towards understanding that the availability of labour was caused at least in part by continuing problems in industry – an Admiralty member of the Supply Board had noted that Thornycroft and others had lost naval orders from Argentina, Uruguay, Siam and China to Italian shipyards since July, and the trend of naval orders being placed elsewhere had only become more pronounced since 1932.[53] The major concern was no longer idle capacity – as it was believed the deficiency programme would go some way to addressing this – but more explicitly the issue of skilled labour to actually complete the contracts. The Supply Board's report stated that:

> In this country, foreign orders enable a greater proportion of shipbuilding workers and armament producers to obtain experience of warship construction and equipment; this is a consideration to which the Admiralty attach the utmost importance as unless they can count on the aid of efficient shipbuilding firms, rapid expansion of the Navy in an emergency would be impossible.[54]

---

49 TNA, SUPP3/18, PSO 421, "Note by Chairman of the Supply Board on the Memorandum by Lord Weir," 25 April 1934, 2–3.

50 See, for example, TNA, CAB60/45, PSO(SB)713, "Categories of Staff and Workpeople," 30 November 1937.

51 TNA, CAB60/34, PSO(SB)41, "Computation of Labour Requirements for Industry," 7 February 1929.

52 Cammell Laird, the constructors of *Rodney*, received the contract for the new aircraft carrier in 1935, but had built no other warships between 1925 and 1932 when it received a contract for a light cruiser. Warren, *Steel, Ships and Men*, 254.

53 TNA, SUPP3/18, FCI68, "Italian Competition in Warship Construction – Note by Admiralty Representative," 8 November 1934, 2.

54 *Ibid.*, 2–3.

This was followed with more evidence of the continuing industrial problems and finally with the recommendation in the Supply Board's annual report that the PSOC bring to the attention of the CID's executive committee "the serious shortage of skilled labour ... in the shipbuilding and armament industries in particular, which is likely to prove a major limiting factor [in the expansion of production]."[55] The CID was promptly informed, and action (of a sort) was taken: the Ministry of Labour was asked for input into the problem the same month.[56]

From this, a much more advanced understanding of the nature of skilled labour in shipbuilding was formed in a short space of time. By April 1935, extensive lists of the myriad professions active in naval construction – from anglesmiths to boilermakers, carpenters, electrical fitters, gear cutters, riveters, welders and dozens of others – were listed alongside estimates of their current numbers, how these matched with anticipated future requirements, and the relative claims on each that the individual SCs felt they might need.[57]

These calculations uncovered the scale of the problem: while unemployment remained high, the total number of engineers and other skilled staff theoretically available fell far below the level the three defence departments believed would be needed if all were simultaneously working at maximum capacity. Moreover, many of the engineers currently employed worked in automobile factories, which it was anticipated would be turned over to war production in the event of conflict but were not intended to be used for rearmament before that point.[58] Weir and Lithgow advocated some general de-skilling of tasks where possible, but since both they and the PSOC anticipated that the full cooperation of trade unions would be essential to maximise efficiency, they believed that antagonising them by replacing skilled with semi-skilled or unskilled workers was not immediately desirable, and such a view prevailed throughout 1935 and 1936.[59]

While these further investigations, still under the veil of secrecy, achieved a far greater degree of nuance and detail than had hitherto been the case, the underlying problem was still that if an expeditionary force on the continent required vast quantities of small arms and field artillery, the composition of skilled workers necessary and the demands placed upon them would shift accordingly. This was, quite obviously, an unsatisfactory situation. However, not much could be done to train new staff or sharpen skills in engineering without more orders being

55 TNA, SUPP3/18, "Note on the Seventh Annual Report of the Supply Board," 10 November 1934.

56 TNA, SUPP3/18, letter to Myrddin Evans, Ministry of Labour (no author given), 12 November 1934; and GUAS, DC35/31, "Twelfth Annual Report of the Principal Supply Officers Committee," November 1934.

57 TNA, SUPP3/18 and CAB60/45, PSO(SB)591, "Emergency Labour Requirements for Naval Shipbuilding and Marine Engineering," 28 April 1935.

58 *Ibid.*

59 See, for example, CCC, WEIR17, 20th meeting of the Subcommittee of Defence Policy and Requirements," 7 May 1936.

placed, educational or otherwise (or indeed winning naval contracts from foreign governments), none of which appeared likely in the short term.

Japanese Naval Rivalry and Spending, 1934/1935

The Supply Board's attempts to calculate the availability of skilled labour had tackled one issue but uncovered another. Knowing where labour was required and what quantities would be needed was directly dependent upon the scale of the construction programmes and their proposed dates for completion. The aerial aspect of the deficiency programme was revised during these calculations and brought forward by two years, but the naval aspect was arguably more pressing yet still unanswered. The Washington and London treaties of 1922 and 1930 were due to expire at the end of 1936, leaving potentially unrestricted naval competition by former treaty signatories. Britain had little desire to enter another naval race to keep pace with its rivals to maintain a meaningful presence around the globe, but without an extended agreement such a course may have been inevitable. Yet without knowing one way or the other, construction levels – and thus labour calculations – were moot.

To seek answers, informal discussions were held with the United States and Japan in the latter third of 1934. Old disputes left over from the London conference reared up once more, and no side could agree upon levels acceptable to the others. Japan now demanded parity in tonnage (up from the 10:10:7 ratio for cruisers and smaller ships agreed with Britain and the U.S. in 1930), with freedom within that limit to build whatever type of vessel she so desired. In addition, France and Italy also sought upward revisions to the existing terms. These profound differences could not be reconciled, and in December Japan announced its intention to let its treaty obligations expire at the end of 1936.[60]

This was an even larger problem for Britain than it first appears. Japan had negotiated treaty limits well above its current levels in both 1922 and 1930, leaving it the most "room" to build up to agreed tonnages. Thus, it had managed to increase spending by ninety percent across the preceding five years (Britain's had fallen by twelve percent) without contravening the Treaty.[61] In other words, the Japanese had considerable freedom to expand its navy while Britain was forced to cope with the fleet it had against an increasing Japanese threat in the Far East. From 1936, that problem was only going to get larger, for even if the domestic situation allowed for immediate orders of capital ships, many would take three to five years to be combat-ready.[62]

---

60 Gibbs, *Grand Strategy,* 156.

61 Hansard, HC Deb, vol. 299, c. 600, 14 March 1935.

62 Rearmament construction times are taken from GUAS, UCS1/1/12, "minutes of Board Meeting at Brown, Clydebank," 17 February 1938.

In this environment, the Admiralty informed the Foreign Office that it believed that the best course of action would be to explore limitation options with France and Germany to free European resources to be deployed in the Far East.[63] Thus, in early 1935 Admiralty and Foreign Office figures discussed the option of a pact with Nazi Germany, which culminated in talks with Hitler in March and April, three weeks after the publication of the White Paper on Defence.[64] This move to discuss separate naval arms limitations with Germany, a non-Treaty power, was a clear step away from the multilateral agreements that had held disarmament together since 1922 and risked antagonising France, Italy or both. In other words, the document that had restricted a naval arms race for the past thirteen years was now hanging by a thread.

---

63 TNA, CAB23/81/12, CC 12(35), Cabinet Conclusions, 27 February 1935.

64 Gibbs, *Grand Strategy,* 164–165; and Cmd. 4930, *Exchange of Notes Between His Majesty's Government in the United Kingdom and the German Government Regarding the Limitation of Naval Armaments*, June 1935.

Chapter Eight

The White Papers, 1935–1936

The *Statement Relating to Defence* (more commonly known as the White Paper) of March 1935 was a watershed in the history of defence between the wars, at least insofar as it finally put rearmament and deficiencies into the public domain. In terms of the background to its publication, there had been little movement from the National Government since Manchuria towards publicly announcing an intention to rearm. Although the Admiralty had constantly warned of the Japanese threat from the second half of the 1920s, the CID had harboured deep concerns about the Far East and Germany as far back as 1930, and the PSOC had highlighted the supply and labour bottlenecks connected with future defence schemes from 1933, the National Government had remained outwardly reluctant to support rearmament or industrial mobilisation. This position did not change in 1934. The reason for this is not hard to locate: public support for any form of defence spending had been extremely low since the end of the Great War, and any hint of armed conflict thereafter usually resulted in some sort of outcry against British involvement. Although there was a growing feeling of insecurity about developments in Germany, this had not managed to overcome the strong pacifist sentiments that still existed. At best, the German situation had reaffirmed public support for the concept of "collective security," primarily through the League of Nations, but until that point events such as Manchuria and the talks in Geneva had only served to highlight to the British government just how strong was the distaste for war. Thus, this was a period where deficiencies were not hard to spot, but nevertheless they were not remedied.

Politics, Pacifism and "Going Public," November 1934–March 1935

The pacifist sentiment is borne out by almost all polling between Manchuria and the first White Paper. "Peace candidates" had performed particularly strongly in by-elections between October 1933 and December 1934 – in Fulham the National Government majority of more than 14,500 was replaced by a Labour majority of nearly 5000, while in Lambeth and Putney the swing away from the National Government was even higher. A straw-poll in a London newspaper had more than seventy-five percent of respondents desiring Britain to stay out of any Franco-German conflict, regardless of the circumstances. Moreover, almost 12,000,000 people signed the League of Nations Union's so-called "Peace Ballot" from 1934, with eight in every ten supporting the total abolition of military aircraft and more than nine in ten supporting ongoing international disarmament.[1]

Quite apart from the ideological criticism of armaments spending, the economy was recovering but was not at full strength. Unemployment remained high – twelve percent in 1934, compared with seven percent before the Depression. Thus, popular sentiment indicated that Britain's first priorities should be the restoration of prosperity at home, not international affairs. MacDonald thus believed that any significant defence expenditure ran the risk of a "financial smash" and urged restraint.[2] Guiding his belief was earlier National Government policy: unemployment benefits were cut under the terms that brought his coalition into existence, and the British public expected this to be reversed once the economic outlook improved.[3] As the National Government was seeking re-election in 1935, it understood that unless – or perhaps even if – it delivered the economic stability it had promised in 1931, the Labour vote would rebound very strongly, as Fulham indicated.[4] These were powerful messages to MacDonald, Baldwin and Chamberlain. Taken together, they served as further constraints on the National Government's freedom of manoeuvre.

That said, from November 1934 there was a more concerted effort to use the anxiety over ongoing German rearmament to educate the electorate of the problems of British defensive deficiencies. In October, Foreign Office intelligence had pointed to a German air force that was expanding far faster than first anticipated, an expansion that dwarfed comparable British plans.[5] The pressure to respond prompted both Lord Londonderry (the Secretary of State for Air) and Baldwin to stress publicly that while Britain was *not* intending to rearm, it did need to "make good" on its shortfall of equipment for defence, with explicit reference to German re-armament and the Nazi party's refusal to cooperate with the Geneva Disarmament process. Linking the two, Baldwin on 28 November announced an increase in the pace of construction of aircraft and aerial defences, with the completion date of the air programme now being set at 1937. This was patently not to be able to fight a war against Germany, but merely to safeguard the British Isles from attack (and to head off some of the criticism that followed from the Labour party).[6] While this solved one problem, it created another: by

1 Kyba, *Covenants without the Sword*, 99, 119–120 and 149.

2 TNA, CAB2/6, "Minutes of Committee of Imperial Defence," 22 November 1934, 1.

3 Morgan, *Ramsay MacDonald*, 79; and N. Smart, *The National Government, 1931–1940* (London, 1999), 108–109.

4 J.C. Robertson, "The British General Election of 1935," *Journal of Contemporary History*, 19/1 (1974), 150–151.

5 Gibbs, *Grand Strategy*, 135.

6 Hansard, HC Deb, 28 November 1934.

linking Germany and air, it made it more difficult to announce any new army or navy spending, as neither could be justified on the same basis.

The senior members of the National Government thus trod a very fine line. The extant threats had to be made clear to justify some remedial action, but not in such a way that would erode already fragile public support or convince the Nazi leadership that Britain desired head-on conflict. The pressure was telling: by this point, MacDonald had long been in poor health, and it was widely believed he would soon be forced into retirement.[7] As such, from the latter part of 1934 spokesperson duties for the government fell to Baldwin, and it stayed this way. During a parliamentary debate, Baldwin sought to soothe the mood by reiterating his belief in arms limitations and downplaying the more alarmist predictions – most notably from Winston Churchill – that by the end of 1937 Germany would have an air force double the size of Britain's. Baldwin conceded that while German preparations were a cause for anxiety, a limited scheme to ensure British defensive superiority was all that was required for the time being.[8] Politically, this achieved its goals in November 1934, but it was hardly a conclusive step towards meeting the German threat or a clear sign to Hitler that Britain desired peace. Nor could this line of reasoning hope to broach the subject of any future naval programmes; Germany remained a comparatively weak naval power that was unlikely to be a threat to Britain for years to come. As such, Baldwin's speech was counter-productive: some in Germany believed that Baldwin had just given tacit consent to Nazi rearmament plans, and the shares of the publicly-listed German private armament manufacturers rose in Berlin.[9] It was also a decidedly mixed message to the private armament manufacturers who were still left guessing what direction defence policy would take and what the effect would be on them.[10]

Of course, this was because the National Government had no clear idea of the direction of defence policy. The announcement from Baldwin did not mean that the German threat had become the overriding concern, for what Baldwin said and what the National Government thought were two very different things. Indeed, the German view is not reflected in supply discussions at the PSOC, but rather the opposite. From late 1934 and throughout 1935, the report titled "The Basis on which Supply Preparations are to be Framed," the guiding document based on hypotheses set by the CID, deliberately stated that while the air programme was to be accelerated, requirements should be drawn up for *both* a war in the Far East and a war against Germany in the five years from 1934. In other words, the CID and PSOC were, quite rationally, still taking the broader view on possible defensive risks and were actively encouraged to plan for what

---

7 Morgan, *Ramsay MacDonald,* 80–81.

8 Hansard, HC Deb, 28 November 1934.

9 Gibbs, Grand Strategy, 141, 166.

10 See, for example, GUAS, GD319/12/7/6, Crease to WSBC Committee, 9 October 1934.

would be needed in the first twelve months of either kind of conflict.[11] These were hampered by uncertainty over future naval limitation agreements, meaning that specifics could not be addressed, but this reflected the geopolitical situation and treaty complexities more than the failings of the subcommittees. Therefore, to all intents and purposes the same problems existed in 1935 as in the immediate aftermath of Manchuria, so planning rather than spending had to continue for the time being. There was some will to address deficiencies and meet the shortfalls of labour and equipment, but not so much that the National Government would risk going to the public and announcing a programme to be paid with further austerity measures in welfare. One can argue, however, that the time was now ripe for action, and the White Paper that followed was the ideal moment to outline an expansion of industry to ease anticipated bottlenecks.

The First White Paper on Defence and Anglo-German Naval Pact, 1935

Instead of pressing on with industrial expansion, these continuing political constraints explain the rather limited nature of the White Paper, known officially as the *Statement Relating to Defence*, published on 4 March 1935.[12] It had been drafted in January and February by the same group of civil servants that had been responsible for the DRC investigations between November 1933 and February 1934, and was to all intents a continuation of the National Government's attempts to educate the public of the growing external threats to Britain's security. Even Hankey, one of the strongest advocates of rearmament – especially naval rearmament – felt that public opinion was not yet ready to accept the privately-held views of the CID and the Cabinet.[13] Thus, there was a very deliberate effort in the paper to highlight deficiencies in Britain's "first line of defence" rather than deficiencies in offensive capabilities, and to show how these deficiencies prevented Britain from playing a full role in a system of collective security. It did not attempt to conclusively answer the question of what needed to be done. By the same token, there was a clear and sustained attempt to link British armaments with concepts of peace and deterrence rather than as preparations for war.[14]

Hankey (and the DRC's) influence was good for the Admiralty, as the document stressed the central role of the navy in British defence. In affording priority to the protection of British citizens through deterrence, the paper affirmed that while the "establishment of peace on a permanent footing" was the

---

[11] TNA, SUPP3/18, CIC1171-B, "Basis on which Supply Preparations are to be Framed," April 1935.

[12] *Statement Relating to Defence*, March 1935.

[13] CCC, HNKY7, Hankey to MacDonald, 10 January 1935.

[14] *Statement Relating to Defence*, March 1935.

"principal aim of British foreign policy," it also claimed that "if peace should be broken, the navy is, as always, the first line of defence." Thus, the navy was a defensive force, essential for safeguarding British interests in an uncertain world. A potential war in the Far East was deliberately omitted (despite being the main concern of the DRC report), while the communications, raw materials and food supply that Britain, as an island nation, relied on for its very existence were repeatedly stressed. The Empire was placed firmly in the context of British security: Britain needed to defend its imperial possessions, including its network of "bases, fuelling stations and harbours" to protect the merchant vessels on which the "British way of life depended." Following from this, the paper argued that unless the long-term security of Britain's ports and sea passages could be ensured, "all other measures of defence are of little avail," and despite developments in the power and range of air forces, British merchant ships delivering supplies "remained as open to naval attack as ever." For these reasons, the CID stressed it was not seeking to expand the fleet – government policy remained absolutely committed to the limitations put in place at Washington and London – but the battleship remained "the essential element of upon which the whole structure of our naval strategy depends." Thus, the navy required ship replacement and modernisation solely due to the advanced age of a high proportion of British vessels. Japan, incidentally, was only mentioned in the context of its participation in the upcoming naval conference of 1935.[15]

There thus was a considerable gulf, even in March 1935, between what the DRC, CID and increasingly the Cabinet believed were serious threats and what the National Government felt able to announce the public and wider world. Having argued in private that Germany was a potential enemy in the "five years from 1934" and that Japan was an immediate concern, the White Paper played up the threat on Britain's doorstep and played down entirely the problems in the Far East. But in terms of priorities the White Paper still looked much more like the DRC's first report than Chamberlain's (re)conceptualisation. It was a decidedly mixed bag. Nevertheless, the statement was still savagely attacked by the remainder of the non-National Labour and Liberal parties as a "complete change of policy' that would lead Britain away from peace and backwards into the "anarchic world that brought us the war of 1914–18."[16]

In terms of industry, the White Paper was symptomatic of a wider problem in National Government thinking which was clearly torn between national security – albeit different definitions of it – and both economic stability and the public mood. As tentative as the White Paper was, it still rested on the foundation that rearmament meant new orders for new weapons which cost far more than the government was prepared to commit at this stage. The deficiency programmes had come along at the same time as a modest revival in merchant shipbuilding,

---

15 *Ibid.*

16 Hansard, HC Deb, vo. 299, c. 46, 11 March 1935.

but this was insufficient to revive the armaments industry. For instance, Yarrow received just two Admiralty orders (for small destroyers) between Manchuria in 1931 and Abyssinia four years later, while Cammell Laird launched just 50,000 tons of vessels in total in the same period, with the only notable Admiralty order being a small cruiser.[17] The work-sharing terms of the WSBC meant that this level of construction was typical of the rest of the sector.

The underlying issue was that the National Government did not separate the industrial base on which any rearmament effort was to be founded from the product, be that ships, tanks or aeroplanes. It understood the industrial problems but was not seriously thinking about solving them unless it was part of a broader rearmament policy. The PSOC – and Weir, Lithgow and Balfour – was thinking completely differently. The government continued to pay little attention to industry and to leave it in limbo, despite recommendations from the Supply Board and Advisory Panel to place educational orders or to introduce schemes to strengthen the facilities and skills base of armament manufacturers. The lack of foresight is most aptly illustrated in the actions of Sir John Jarvis, a businessman who first raised £40,000 through donations and then spent a further £100,000 of his own money to assist the unemployed shipbuilders of the Palmer yard in Jarrow by, among other things, buying the *Olympic,* the retired sister ship of the *Titanic,* and bringing it to Tyneside for scrapping.[18]

The Admiralty had ideas to assist industry to maintain skills and capacity, but these were typically limited. Chatfield's armour subsidy that had been brought down under the weight of Treasury opposition in 1929 was re-started with Treasury consent, albeit as part of an agreement to pay a higher price per ton of armour if the three major firms (Beardmore, Brown and Vickers) committed to retain capacity capable of producing up to 18,000 tons per annum between them. While not a bad start, this was only a fraction of what would later be needed. Other schemes in 1934 included diverting money from research and development to construction, which paid for the new engines that were fitted to HMS *Warspite* that year, with a price fixed by the WSBC.[19]

Events, however, overtook the limited aims of the White Paper. Within weeks, Hitler stated his intention to produce enough aircraft to achieve parity with France (which possessed a far larger air force than Britain), which meant that the estimates of German air power had to be revised upwards. In response, the DRC and the ministerial committee from 1934 were re-formed in the summer of 1935, while a committee on the expansion of air power was formed with Lord Weir as a full member.[20] At the same point, talks to bring all major naval powers

---

[17] Borthwick, *Yarrow and Company Limited*, 54; and Warren, *Steel, Ships and Men,* 253–254.

[18] M. Perry, *The Jarrow Crusade: Protest and Legend* (Sunderland, 2005), 131.

[19] Gordon, *British Sea Power,* 85; Hume and Moss, *Beardmore,* 240; and TNA, ADM167/90/3182, Board of the Admiralty, minutes, 15 March 1934.

[20] CCC, WEIR19, Lord Londonderry (Secretary of State for Air) to Weir, 20 May 1935.

to a new conference in London before December were underway. Britain had also extracted an offer – necessarily secret as not to alarm France – from Hitler to limit the German navy to thirty-five percent of the British level. This was accepted on the basis that Hitler was probably going to build a fleet regardless, and as such it was best to accept a binding offer on the table.[21] The pact was finalised in June 1935, and for the time being smoothed over one part of the naval problem, although like the Washington and London treaties before it, effectively allowed the tiny German navy several years of unrestricted construction to "catch up" to its agreed limit.[22]

Despite the failures to address the fundamental material weaknesses in British defence with the White Paper, there were some positive naval developments. A side effect of the heightening of the aerial threat and apparent reduction in naval antagonism was that scholarly analysis of the period surrounding the White Paper has been much heavier on air programmes than sea power. As such it has been used by historians to illustrate the increasing centrality of the RAF at the expense of the navy and army,[23] but the reality was far more nuanced. It is clear that while the revisions to the aerial programme were significant, the subsequent Anglo-German naval pact and second London naval conference have perhaps misled some observers. The White Paper played a key role in publicising for the first time between the wars the intention to rearm while placing the role of the navy centrally within that process. This point has been overlooked in favour of focusing on the aerial threat. In this sense, the White Paper was a limited document, but one that nevertheless reflected the work of the CID's subcommittees since Manchuria in highlighting the need for increased naval defence and new naval construction. Like the CoS report, it believed that a naval threat required a naval solution, and like the PSOC and Supply Board, that the "deliberate retardation" of armaments since Washington was a grave cause for concern. In sum, the work of the PSOC and Supply Committees between 1932 and 1935 had been translated into the White Paper's contents.

Still, one should not be too quick to assume that the PSOC was having a profound impact on defence policy. The major stumbling block was not in getting Cabinet to accept that supply problems existed but in getting Cabinet to take steps to at least partially alleviate them, even from within its constrained position. For all the work of the various subcommittees in the previous three years, there was little progress in this regard.

---

21 Cmd. 4930, *Exchange of Notes Between His Majesty's Government in the United Kingdom and the German Government Regarding the Limitation of Naval Armaments*, June 1935; and Gibbs, *Grand Strategy*, 164–165.

22 Once finalised, Britain did have to inform the Americans, Japanese and French, however. The former two powers had no problems with the arrangement, although France protested loudly.

23 Bell, *Chamberlain, Germany and Japan,* 124; and Neilson, "Defence Requirements Subcommittee," 678.

Industrialists, the Second White Paper and Second Naval Conference, 1935–1936

Although the increase in armament spending and the necessity of sea defence had come into the public domain from the end of 1934, it was not "rearmament" that the electorate was being sold. Major rearmament plans only followed the Italian invasion of Abyssinia and the general election from October 1935. Between the Anglo-German naval pact and that point, the DRC was reconvened to re-examine the overall condition of British defences and the steps to be taken considering recent international developments. Due to the pressing nature and fluidity of the international situation, the DRC prepared its next report in just three weeks.[24] This, like the first, stressed the continuing problems in the Far East and the possibility of an alliance between Germany and Japan.[25] In addition, it noted that defence spending in Germany had reached almost one billion pounds annually (the comparable British figure was under £150m), that Japan's naval strength was increasing relative to Britain, and the potential problems if Italy did not remain neutral.

The crux of the report was not so much a change in time frames – it still considered war unlikely before 1939 – but how any rearmament was to be paid for. For Britain to equip adequately for its own security and to discourage aggression, the National Government was told there was "no alternative" but to "widen its horizon and resort to some system of capital expenditure."[26] In other words, tax more, spend less or take out another loan to finance rearmament. That this idea was signed off upon by the Permanent Secretary of the Treasury (Warren Fisher) as part of the DRC indicates at least a partial relaxation of financial stringency on his part – but this should not be taken as evidence that the Treasury, or indeed the Cabinet, had subscribed to any of the tax, cutback or loan options. It was, however, markedly different from MacDonald's warning of a "financial smash" only a few months earlier.

Like the first report, this was to be put in front of ministers on an interim basis before being sent back to the CID with the final agreed-upon guidelines. There was, however, one major difference. In 1934, the group of ministers reviewing the first DRC report was known as the "Ministerial Committee on Disarmament," while in April 1935 it was formally reconstituted as the Defence Policy and Requirements Subcommittee (DPR). This committee (despite being sometimes incorrectly listed as a subordinate body of the CID) from this point onwards sat atop the organisational structure as the Cabinet organ which considered and acted upon the work emanating from the CID.[27] Its name alone suggests an important

24 Gibbs, *Grand Strategy,* 178–179.

25 TNA, CAB16/109, DRC25, 2nd Report of the DRC, 29 July 1935.

26 *Ibid.*

27 Gibbs, *Grand Strategy,* 175, 180.

change in emphasis, and indeed in many ways marked the end, in planning terms, of the deficiency programmes and the starting point of rearmament.

The DPR's membership for the first meeting was drawn from the very top echelons of politics – the Prime Minister took the chair, while the Lord President of the Council, Chancellor of the Exchequer, Foreign Secretary, War Secretary, First Lord, Air Secretary and President of the Board of Trade comprised the rest of the committee. The CoS, Hankey (who was also Cabinet Secretary), and the Treasury and Foreign Office permanent secretaries also attended, although they were not listed as full members of the committee.[28] Since the DPR was now notionally the top level of the CID pyramid, its actions are important, for they reveal much about the nature of defence and supply planning – and its successes and failures.

The first meeting in late July concerned only the second DRC report. Thereafter, the DPR met on a regular basis, beginning in October. At the first meeting the DPR agreed with all recommendations but still expressed caution in moving immediately towards rearmament. Instead, the DRC was instructed to proceed along the lines of the interim report and to prepare full defence programmes for all three services on the assumption that would be complete and paid for by the conclusion of the 1938/1939 financial year. In industrial terms, the DRC was asked to suggest what "special measures would be required for increasing factory output to provide the material required" and "how long a period would be required to reach a state of preparedness." This, of course, was still with the caveat that "existing financial and political considerations" would have to be considered.[29] On one level, this was just more thinking and more planning, but on another it marked an important, if tentative, step towards a coherent rearmament policy.

The DRC, in conjunction with other CID subcommittees, set to work between August and November. Owing to the upcoming general election and the status of the DPR's members, relatively little else was achieved before December, as the committee's senior Cabinet members set to work on the National Government's re-election bid. As had been predicted, MacDonald stepped down as Prime Minister in June, citing failing health, and was replaced by Baldwin.[30] MacDonald remained part of the multi-party National Government, swapping places with Baldwin as Lord President of the Council, and together the government returned a sizable, but substantially reduced, majority.[31]

In the run-up to the election, over the summer the deteriorating Abyssinian situation – first sparked by border clashes with Ethiopians in Italian Somaliland

---

[28] CCC, WEIR17, DPR(DR)1, "Committee on Defence Policy and Requirements – Terms of Reference on Defence Requirements," DPR(DR)1, 30 December 1935.

[29] TNA, CAB16/109, 4th meeting of the DPR, 31 July 1935.

[30] Morgan, *Ramsay MacDonald,* 81.

[31] Smart, *National Government,* 119.

in December 1934 – had overshadowed the normal course of campaigning, although it did not become a major feature of future defence planning in the CID until August.[32] Throughout the crisis, the League of Nations had repeatedly failed to temper Italy, and on 3 October 1935, exactly six weeks before the election, Italian forces invaded Abyssinia, sparking an eight-month war. Further League attempts at sanctions against Italy were similarly ineffective – severely damaging the organisation's credibility – and the National Government's inability to decisively influence the situation was an embarrassment to Baldwin and his Cabinet.[33] The net result in Britain was a sharp swing in public opinion towards a stronger British stance and some form of rearmament.

For these reasons, historians have debated the role that armaments policy (and the notion of collective security) played in the result of the election, and whether the vote delivered a mandate for rearmament or if the public had been misled by the National Government's rhetoric. Thus, a brief examination of the campaigning around the general election is relevant at this juncture. In many ways, campaign rhetoric was not much different than the White Paper which preceded it in that what the National Government believed was far removed from what it was willing to say. Less than one-quarter of the fifty or so different official election leaflets from the government mentioned defence or international relations; most of the focus throughout October and November was on the economy and the domestic situation.[34] Similarly, foreign affairs were mentioned only briefly in speeches by senior government members. The clauses in the manifesto which referenced defence read much like the White Paper of eight months earlier. Collective security and the prevention of war was the "keystone" of government policy, while the defence programme would "be strictly confined to what is required to make the country and the Empire safe, and to fulfil obligations towards the League."[35]

Considering the experiences not only of the previous twelve months but also of any time since the Wall Street Crash and the onset of the financial crisis, such rhetoric should not be particularly surprising. In by-elections or polls where the National Government fared poorly, analysis indicated that the economy and welfare had a far higher priority than the international situation.[36] In this sense, Baldwin and his colleagues were behaving as rational and calculating politicians, but in not reinforcing the seriousness of defence issues, the National Government opened itself up to the charge of misleading the electorate when rearmament soon followed. Support for the League had been wavering within the Cabinet

32 For overview of progress and early recommendations, see CCC, WEIR17, 6th Meeting of the DPR, 5 September 1936.

33 Robertson, "The British General Election of 1935," 159.

34 *Ibid.*, 159–160, 164.

35 I. Dale, *Conservative Party General Election Manifestos, 1900–1997* (London, 1999), 53–59.

36 Kyba, *Covenants without the Sword,* 99–100.

for a significant period before Abyssinia, and in hindsight the commitment to it appeared insincere. Moreover, the repetition of the claim that spending would only be to meet deficiencies was a missed opportunity. The League's credibility had been shaken, the international situation had further deteriorated, and the chances of a National Government majority were exceedingly high. Yet the rhetoric was no stronger than it was eight months earlier in March. As such, Britain lost more ground.

The first major act of the re-elected government was to summon the DPR to examine the DRC's third report, which was nearing completion in November, and to convene the Second London Naval conference, which began in December.[37] Preparations for the latter highlighted Britain's relative decline in naval capabilities. Referencing Japan's recent naval increases, the First Sea Lord Ernle Chatfield (at this point also working on the DRC report showing Britain's levels of industrial preparedness), urged the limitation of the size of ships and guns constructed, even if a general limit on ship numbers could not be agreed. On the industrial situation, Chatfield believed that failure to implement size limitations would allow Britain's rivals to

> … spring a surprise which is exceedingly difficult to meet, because the designing or new guns and mountings takes a number of years and once a country has got a start by secret preparations her opponent cannot catch up to her quickly.[38]

In other words, because it did not build new ships to replace older ones, Britain was potentially not developing new technology – and not sharpening technical skills – at the same rate as other nations that had been allowed to build up to treaty limits over the previous five years. It would therefore be vulnerable if a rival suddenly developed larger and better guns.

Chatfield's point was borne out by the composition of the fleet. In December 1935, the most modern British ships with large guns were the *Nelson* and *Rodney* battleships, essentially cut-down versions (to meet Washington Treaty limits) of late WWI designs.[39] The others were remnants of the last war. Of the smaller post-war heavy cruisers, the *County* and *York* classes were only around one-quarter of the displacement of the battleships and were built to Washington limits, meaning to meet total weight restrictions they lacked sufficient armour, were equipped with much smaller weaponry, and were based on designs that were by then over a decade old. In terms of construction in the five-and-a-half years since the London Treaty, sixteen light cruisers – around one-sixth the size of *Nelson* – had been laid down, of which eight were in commission. But these

[37] Roskill, *Naval Policy between the Wars*, II, 314.

[38] TNA, CAB29/148, NCM (35)23, Chatfield memorandum, 31 October 1935.

[39] Buxton and Johnston, *Battleship Builders*, 35.

were of limited combat use even for trade protection and were certainly inferior to heavy cruisers and no match for battleships. In terms of the large and heavy naval guns, which the Third Sea Lord had called "invariably the limiting factor in the construction of every kind of warship"[40] in 1933, just one substantial order – in 1923 for *Nelson* and *Rodney*'s weapons – had been placed in the thirteen years since Washington in British private yards.

By way of comparison, the Imperial Japanese Navy had launched six heavy cruisers in the same period and had four more under construction. These ranged from 11,000 to over 15,000 tons and were significantly larger and more powerful than the equivalent light British cruisers, at the same time from a newer generation of vessel than their ageing heavy British counterparts.[41] This aptly highlights the failings of the treaties in practice. For while in principle the idea of arms limitation between 1918 and 1930 was sound, Britain's position as the largest navy meant she had most to lose; despite being "limited" to a fleet that was sixty to seventy percent the size of the Royal Navy, Japan significantly out-built Britain in some key areas over a period of several years, produced larger ships with bigger guns, maintained facilities and skills, and then withdrew from the treaty when it expired. Although the sixteen orders for light cruisers since 1930 – seven as part of the DRC scheme since mid-1934 – were a welcome boost to the Royal Navy, just nine were offered to the private yards for tender.[42] The lack of large ship orders since 1922 cannot be separated from the previously expressed concerns by the Supply Board and the Admiralty about the lack of remaining capacity for the construction of large guns. Again, the difference between how the lower levels of the PSOC conceptualised the supply problem and top-level policy was marked.

As had been expected, Japan could not be persuaded to sign the Second London Treaty. Neither was an agreement with France, Italy or the United States possible with regards to total tonnage limits and ratios. After months of negotiations, an agreement in principle was reached in March 1936 with the other naval powers to limit the size of guns to fourteen inches (eight inches on cruisers), with an allowance to raise the limit to sixteen inches should any of the former signatories to the 1922 treaty fail to adhere to these new restrictions in gun calibres.[43] As such, the era of quantitative treaty restrictions, which had helped limit British naval construction across the previous fourteen years, was brought to an end – although because the 1930 treaty was in force until the end of the year, no new work could commence immediately.[44]

---

[40] TNA, CAB60/30, PSO(SB), 24th meeting of the Supply Board, 11 October 1932.

[41] McMurtie (ed.), *Jane's Fighting Ships*, 287–295.

[42] Derived from *ibid.*, 18–41. Japan had some 84,000 tons of cruisers in active service by January 1936 – 14,000 more than Britain.

[43] In practice, the fourteen-inch large gun limit was followed only by Britain. Japan immediately broke the clause and laid down ships with larger guns, so the United States and Italy followed suit.

[44] Cmd. 3758, *Treaty of London of April 22, 1930*, April 1930.

The signing of the treaty was contemporaneous with the second *Statement Relating to Defence* (White Paper) that had been published three weeks earlier.[45] Like the previous paper, the 1936 statement was based largely on information provided by the DRC, this time in its third report in late November 1935.[46] Again, there is something of a parallel narrative here, with the discussions of the Second London Naval conference taking place at almost exactly the same time as the DPR's examination of the third report. The content of the DRC report was a closely guarded secret, with very few members of the Cabinet even privy to its contents.[47] The report made it clear that the original deficiency programmes had not gone anywhere near far enough and that a three-front war with Italy, Japan and Germany was a possibility. Moreover, German naval rearmament, coupled with Japan's refusal to agree to a lower tonnage limitation than Britain, left no margin for security should both become hostile, especially given a likely commitment to the Mediterranean to face an Italian threat. Thus, the report urged a new standard of naval strength whereby a permanent deterrent force could be placed in the Far East while having a "home" fleet capable of protecting trade and facing the largest European power.[48] Due to the lack of agreements on tonnage limits and ongoing construction in Japan, Germany and Italy, the Admiralty and the DRC felt unable to specify the minimum size, in quantitative terms, of a fleet capable of achieving this. Based on estimates of Japanese and German programmes, the DRC believed that at least seven new battleships, fifteen cruisers and four aircraft carriers would be needed, plus numerous smaller support ships.[49] This was to be known as the DRC's "standard fleet" and would guide naval construction plans from 1936 until 1938.[50]

On industrial preparedness, the DRC painted a grim picture. While it believed the system of supply organisation (the advisory structure with the PSOC and Supply Board) should remain, it noted that so much replacement work was already needed to bring the ageing fleet up to modern standards and that the lack of foreign and domestic orders since Washington had so reduced productive capacity and skills that there was no leeway left to significantly expand the fleet to this level in the three-year time frame.[51] As a result, the DRC believed that industrial capacity, independent of political will and finance, was likely to be a major constraint on the rearmament of the Royal Navy before 1940. This was a marked change from 1934, when the DRC had proposed a

---

45 Gibbs, *Grand Strategy*, 329; and Cmd. 5107, *Statement Relating to Defence*, March 1936.

46 TNA, CAB16/112, DRC37, 3rd Report of the DRC, 21 November 1935.

47 CCC, WEIR17, Hankey to Weir, 6 December 1935.

48 *Statement Relating to Defence*, March 1936.

49 TNA, CAB16/112, DRC37, 3rd Report of the DRC, 21 November 1935.

50 TNA, CAB16/140, DPR88, "Practicability of Accelerating the Naval Programme and the Effect on the Programmes of the Other Services," 25 June 1936.

51 TNA, CAB16/112, DRC37, 3rd Report of the DRC, 21 November 1935.

£21m increase in expenditure.[52] Had more money been spent at that point in the under-utilised yards, it would have had two major advantages. The first is that it would have reduced the amount of replacement work needed in 1936, in effect freeing up more capacity for fleet expansion. The second is that it would have allowed for an increased intake of apprentices and a renewal of ageing facilities, effectively reducing the time lag between beginning rearmament and reaching maximum output. In hindsight, it was undeniably a huge missed opportunity.

In addition, the DRC believed that a "shadow" armaments industry, consisting of automobile and other firms which could be turned over to war production quickly (first proposed by Weir two years earlier) should be initiated. On its own this would still not solve the problems. The DRC believed that about 120,000 extra skilled or semi-skilled workers would be required to meet the requirements of the three services, and rationing of staff among the services would probably also be needed.[53] These points, the last especially, were a direct result of the work the Supply Board had undertaken some months before and showed the National Government the problems of the 1920s coming back to bite.

The industrial nature of the problem prompted a re-think by the DPR. As soon as the ink was dry on the report, the DRC's author, Maurice Hankey, was instructed to write to Lord Weir. He told Weir that:

> The Prime Minister has authorised me to send you a copy of the Report of the Defence Requirements Committee, of which I was chairman, which is now before the government. I must tell you that this Report is of such secrecy that it has not been sent to the Cabinet as a whole, but only to the Committee of the Cabinet which initiated the enquiry. You will therefore treat it as a document of quite extraordinary secrecy.[54]

The "committee" to which Hankey referred was the DPR's subcommittee examining the implications of the third DRC report.[55] Quite apart from a civilian businessman being privy to information that most of the Cabinet had not even seen, there were other striking developments in the pipeline. The Prime Minister wanted the DPR to meet regularly from now on to discuss rearmament and for Lord Weir to be its newest member.

---

52 Gibbs, *Grand Strategy,* 124.

53 TNA, CAB16/112, DRC37, 3rd Report of the DRC, 21 November 1935.

54 CCC, WEIR17, Hankey to Weir, 6 December 1935.

55 This was known as the DPR(DR). Weir also became a full member of the DPR afterwards, though the distinction between the committee and subcommittee is to an extent moot, as in December 1935 they shared memberships and were focused on the same questions. Still, they were recorded as separate committees and have been so classified in the archives. See Appendix 4, below.

Until this point, Weir had been advising the PSOC and serving on the Committee on Air Defence Research.[56] In the wake of the first White Paper, he had been announced in May as an advisor on air deficiencies, and this had led to a flood of letters from well-wishers, almost always former air staff or industrial partners that had worked with Weir in the Great War.[57] Although his work on air defence research is not strictly relevant here, it is worth noting the side effects of his involvement on the occasions when such work was publicly known. Once former colleagues in the Ministry of Munitions found out about Weir's new role, he was inundated with requests to use his influence and connections to "put a good word" in with the Air Ministry about their abilities. So many of these arrived, often from those who had not been in contact with Weir for many years, that his secretary responded with the same template letter, thanking them for their wishes but informing them simply that Lord Weir's role had been "exaggerated."[58]

The inclusion of Weir was agreed upon by the DPR and signed off by its chairman, Stanley Baldwin. Weir, it was decided, "should have discretion in any work he might be willing to undertake on behalf of the Committee" and that he should be able to "associate with Lord Riverdale (but not to consent to Riverdale being a member of the Committee) if Riverdale was prepared to render assistance'.[59] Lord Riverdale, of course, was Sir Arthur Balfour's title from 1935.[60] In addition, Weir was instructed to

> … turn over in his mind, in addition to the [DRC report], the industrial side of our problems … such as the right steps to be taken for organising industry and the production of war material, and the best time and method for approaching both employers and labour with a view to securing their effective cooperation …[61]

Essentially, this meant that Lord Weir was to be given significant latitude to develop plans to organise the industrial side of any future armament programmes, including the method of approaching and co-opting industry, with the assistance of Balfour. Conversely, the DRC members – which included the heads of the three defence departments – were permitted to "attend the Committee as advisers … except when the members preferred to meet alone."[62] Thus, in this structure, Weir

---

56 TNA, CAB16/132, ADR1, 1st meeting of the Subcommittee on Air Defence Research, 10 April 1935.

57 See GUAS, DC96/21/1–2.

58 For example, GUAS, DC96/21/2, Weir to Wheatley, 27 May 1935.

59 CCC, WEIR17, DPR(DR)1, "Committee on Defence Policy and Requirements – Terms of Reference on Defence Requirements," 30 December 1935.

60 Walton, "Balfour, Arthur."

61 CCC, WEIR17, DPR(DR)1, "Committee on Defence Policy and Requirements – Terms of Reference on Defence Requirements," 30 December 1935.

62 *Ibid.*

outranked the heads of the fighting services and was in closer working contact with the PM and Chancellor than any other civilian in the country. Put another way, the most senior government body working on the most secret aspects of national security during the time the government was still outwardly committed to meeting "deficiencies" had a civilian businessman as one of its full members. This, like the Advisory Panel before it, was not only a peacetime first but also represented a significant increase in Weir's access to information, as well as his responsibilities and influence.

Weir immediately played a full and active role on the committee. About the Third DRC report, he believed that the industrial position was worrying and that Britain should concentrate on building a deterrent force first, at the expense of Army programmes if need be.[63] This was supported by Chamberlain, but not everyone was impressed. Hankey believed that a small and efficient army would also be an effective deterrent, while Henry Pownall, the Deputy Secretary of the CID, remarked privately afterwards that Weir's ideas were "absurd" and that that Weir's belief in the inability of private industry to carry out the defence programme was because of his "innate Scottish caution."[64] Nevertheless, Weir and Chamberlain succeeded in reducing the Army demands listed in the third DRC report, and Weir was asked to continue with his work on the state of industry.[65]

In January, Weir submitted a full industrial report to the DPR. This in many ways was an update of the Advisory Panel report of twenty-one months earlier, with similar advice on the expansion of supply and whom to approach in industry. However, there were some notable new aspects. On skilled labour, he warned that "securing reasonable efficiency … to avoid the misuse of existing skilled labour" was a "major trouble that could seriously affect progress," but counselled that the CID should avoid any broad consultation with labour unions for the time being, as it would create a counterproductive "crisis-type" atmosphere. Instead, he felt that the committee should instead tackle problems as they arose. Most importantly, on the question of authorising the financing of defence orders, Weir suggested the following:

> To achieve [the goals of the DRC report], the Supply Departments and contractors will be working against time. Decisions instructing new contractors to proceed … must inevitably be given by the Supply Staff prior to the settlement of financial details. In this sense … the word of the man responsible for Supply must carry, and the spirit and enthusiasm he has evoked in the contractor's mind must not be chilled by delays of approvals caused by financial control.[66]

63 B. Bond, *British Military Policy between the Two World Wars* (Oxford, 1980), 223.

64 TNA, CAB21/422, Note by Henry Pownall, 13 January 1936.

65 Bond, *British Military Policy between the Two World Wars*, 224.

66 TNA, CAB16/123, "Subcommittee on Defence Policy and Requirements, Programmes of the Defence Services. Report," February 1936, 5.

This recommendation was aimed squarely at the Treasury and was undoubtedly shaped by his experience on the Advisory Panel in the previous two years, where the ideas and the recommendations of the Supply Board and his industrial colleagues repeatedly stalled while the Treasury examined, or revised, plans for many months at a time. While the Treasury's opposition to extra defence expenditure diminished noticeably in the second half of 1935, it was still causing Weir concern. He finished his report by warning that even if his recommendations were agreed to there was still no way to say whether the defence programmes could be fulfilled, as time was very much against them.[67]

Weir's report was taken very seriously, and the DPR agreed that he should continue investigating the subject. In the meantime, Cabinet drew up and published the Second White Paper. During this time, the German situation was intensifying: the Nazi leadership was publicly flouting the Treaty of Versailles by threatening to occupy (and subsequently occupying) the Rhineland. In addition, the British economy was improving, and public priorities were also shifting noticeably towards demands for the League of Nations to show more teeth.[68] Yet once again, the National Government was reluctant to go public with the full extent of the requirements recommended by the DRC, instead opting to disclose a more limited form of rearmament, again full of justifications for why Britain had been forced into action. It noted that because "conditions in the international field deteriorated" during 1935 and a "deplorable" rearmament process was occurring across the globe, the government felt it was no longer wise to "take risks for peace" as these had plainly not "removed the dangers of war." For these reasons, the government believed it had to take steps, but only those "essential in the present circumstances" to secure Britain from attack. Even these were only being made only after a "prolonged and exhaustive examination" of the condition of the fighting services.[69]

Once again, assistance to the Navy was justified based on preserving essential communications, food and materials, plus free movement between parts of the Empire. But despite the need for a programme of more capital ships and cruisers over several years, the government only detailed the work to be undertaken in the next twelve months, i.e., two battleships, five cruisers and one "smaller-type" aircraft carrier (about one-quarter of the DRC's "minimum" level of construction) so as to not to alarm parliamentary opposition more than

---

[67] Peden, *British Rearmament and the Treasury*, 71; and TNA, CAB16/123, "Subcommittee on Defence Policy and Requirements, Programmes of the Defence Services. Report," February 1936, 5.

[68] Indeed, when the White Paper was presented, even the Labour and Liberal parties no longer criticised it on the principle of arms leading to war, although they did find fault in its "lack of clarity" for the future. Churchill, incidentally, felt the report did not go far enough. See Gibbs, *Grand Strategy*, 267.

[69] *Statement Relating to Defence*, March 1936.

necessary.[70] The paper did not, however, attempt to minimise the seriousness of the problem facing Britain or the difficulty of the industrial challenges. The increased burden on government contractors meant that to "some extent" they might be required to extend their plant or workshops to contribute to the increased output demanded. It was promised that this would not lead to "extravagant profits" for the firms concerned.[71]

On finance, Weir's recommendation was also translated into the report. While it was stated that "Treasury control would be maintained," it was deemed important "that the work is not delayed by the over-elaboration of financial safeguards."[72] Given the role of the Treasury in all defence questions until that point – most notably in the cruiser crisis and the DRC's first report – this clause is remarkable, especially given its origins. This was in many ways the first wresting of financial control away from the Treasury since Washington fourteen years earlier. Overall, despite the cautious language, the Second White Paper conclusively gave the green light to rearmament and was a massive step beyond the rhetoric of deficiencies. It went a long way towards articulating the serious need for imminent expenditure, and was the first public notification to business that there would likely be a prolonged period of military and naval demand and that they should prepare for it. It had been a long time coming, and a small group of industrialists had played a significant role in it.

The Position of Private Industry: James Lithgow, Beardmore and Fairfield, 1935–1936

Between March and July 1936, plans were drawn up for a major naval programme which would ultimately cost tens of millions of pounds and bring contracts, employment and profit to naval armament manufacturers. Before this point the deficiency programmes only represented a moderate improvement in the fortunes of the WSBC firms, which were still in a weak financial position with no guarantee of orders in the longer term. There were just nine contracts for ships larger than 1500 tons placed in private yards between 1930 and the end of 1935. The Admiralty, while in a stronger position to argue for more spending than it had been in the 1920s, had still encountered stiff Cabinet and Treasury opposition. Though the advent of the DRC, a worsening international situation and an improving financial situation had altered the balance somewhat in 1934, it only shifted more conclusively during the last quarter of 1935.

---

70 In the event, these plans were superseded and larger carriers were built. Although no explanation was given, "smaller-type" probably meant a carrier that was closer in size to *Hermes* (11,000 tons) than to *Ark Royal* (22,000 tons).

71 *Statement Relating to Defence*, March 1936.

72 *Ibid.*

The position in private industry between the NSS "mothballing" of yards in mid-1934 and the Second White Paper has not thus far been covered, for there was little overlap between the state and private industry in this period, save for a couple of very significant occasions. This was primarily because from a strictly business point of view there was no outward indication – nor could there be – that long-term rearmament (as opposed to a more-limited deficiency programme) was on the way before 1936. This decision was only taken at Cabinet level around the third DRC report in November, 1935. For this reason, the members of the WSBC were not, even in early 1936, actively preparing for what was to follow. The Admiralty was still tied up with international negotiations over armaments limitation, and as such did not appear to give the shipyards more than a few weeks' notice – still on a provisional basis – of the composition of the 1936 building programme. Indeed, even by September 1936, the WSBC was still attempting to collect confidential information on the number and types of vessels that would be ordered in 1937.[73] In this sense at least, not much had changed since Washington caught the industry by surprise.

For the naval yards, the general outlook for merchant and passenger was somewhat brighter than it was in 1931, but profits remained low, even if the WSBC was successful at bringing in reasonable margins on the few available Admiralty and foreign navy contracts. Fairfield, for instance, still had virtually no margins built into its merchant and civilian contracts into the mid-1930s and was selling them at cost, while the few warship contracts were bringing in ten to twelve percent profit.[74] The PSOC's "educational orders" were similarly slow to materialise, and when they did begin to appear – again from late 1935 – they were almost uniformly aimed at educating some non-specialist armaments makers (and even then only if it was on such a scale that it did not interfere with the firm's "normal activities") in non-naval components, particularly aeroplane frames and engines.[75]

There were nevertheless a few notable differences compared with the mid-1920s. The most significant was obviously the creation of the NSS and the PSOC's Advisory Panel of Industrialists in 1930 and 1933 which brought a shipbuilder, an engineer and an armour maker (but not warship builders) inside the CID framework and gave Lithgow, Weir and Balfour access to important supply information. While the panel obviously had little idea that rearmament would follow before the White Papers, it did help in exposing the nature of industrial deficiencies between 1933 and 1935. As such, the trio had a better idea of what *would* be needed and where the bottlenecks were if the CID's hypotheses

---

73 GUAS, GD319/12/7/6, "Points for Discussion at a Meeting of Firms Capable of Building Capital Ships," 18 September 1936, Things moved rather quickly after that, and the WSBC could plan further into the future.

74 GCC, UCS2/60/1, "Admiralty Orders Completed, 1935."

75 S. Ritchie, *Industry and Air Power: The Expansion of British Aircraft Production, 1935–1941* (London, 1997), 59.

were accurate. Lithgow thus changed NSS policy to mothball warship facilities, and by the end of 1935 Weir and Balfour had been authorised to devise methods to speed up production.

Very few, if any, of these facts were known to WSBC members, in the same way the WSBC's existence was almost certainly not known to the Admiralty. On one side, Weir, Lithgow and Balfour's work was top secret – and the side effects of Weir's work on air parity showed aptly the kind of response when these sorts of connections became more widely known. For this reason, had the panel's connections been understood by the private armaments industry, one would expect the same floods of requests to have been forthcoming. On the other side, the WSBC was still illegally fixing prices (see Table 8.1, below) and risking the wrath of the Treasury, Admiralty and public, so it could not afford word to get out. It is thus unsurprising that the two groups led parallel existences until the middle of the 1930s.

In 1936, however, paths crossed: Lithgow by this point was chairman of both Fairfield and Beardmore, two of the largest armament firms in the country, and (in the case of Fairfield), a member of the WSBC.[76] This was a striking occurrence considering Lithgow's background as a merchant builder – the focus of his entire career to that point. Moreover, there were no other willing buyers for the failing firms – save the NSS and mothballing or liquidation – before 1935. Thus, like Lithgow's involvement with the PSOC and NSS, these acquisitions can be interpreted differently when mapped onto the other events of 1934–1936. It is thus worthwhile deviating briefly from the chronology to examine Lithgow's business ventures.

Until 1934, Lithgow's purchases of companies and facilities were unremarkable. In the 1920s, he and his brother were primarily interested in expansion of the Lithgow site in Port Glasgow, extending the company along the waterfront to the east and west by purchasing parts of other firms to concentrate his business interests in the same spatial area (see Table 8.2, below). In this sense, the purchases after his involvement with the PSOC are a clear break from the past, especially considering the public commitment to disarmament and the lack of interest from would-be rivals.

Indeed, although Lithgow's steel rationalisation schemes in the 1930s are largely part of another story, his acquisition of the Steel Company of Scotland in 1934 was, like the NSS, funded by a Bank of England vehicle. His company had sold its holdings in Dunlop Steel to (or more accurately merged them with) the Colville Steel Company some years earlier, and the new acquisition represented a major rival to Colville for the steel supplied to Clyde shipyards. Moreover, also like the NSS, the Steel Company of Scotland deal was brokered with the help of long-time friend Sir Andrew Duncan, now a director of the Bank of England, who was from January 1935 President of the British Iron and Steel Federation.

[76] Reid, *James Lithgow,* 172.

Table 8.1:
WSBC tendering practices for aircraft carrier order, 1935

Firm	Estimates			Order of tenders	Price to be quoted		
	Hull	Machinery	Total		Hull	Machinery	Total
Cammell Laird	1,395,000	543,000	1,938,000	1	1,487,850	563,125	2,050,975
Hawthorn Leslie	1,238,000	530,000	1,768,000	4	1,551,000	579,000	2,130,000
Fairfield	1,210,000	520,000	1,730,000	2	1,527,000	573,000	2,100,000
Harland &Wolff	1,280,000	535,000	1,815,000	3	1,539,000	578,000	2,117,000
Swan Hunter	1,208,000	525,000	1,733,000	5	1,574,000	571,000	2,145,000
John Brown	1,274,000	540,000	1,814,000	7	1,601,000	574,000	2,175,000
Vickers-Armstrong	1,250,000	540,000	1,790,000	5	1,623,000	575,000	2,198,000
Scotts	1,240,000	556,000	1,790,000	6	1,593,000	572,000	2,165,000
Mean of estimates	1,261,850	536,125					

Source: WSBC Tendering Practices, *Aircraft Carrier, January 1935. (Secret)* GUAS: GD319 12/7/5.

Table 8.2:
James Lithgow's business acquisitions, 1915–1936

Acquisition	Company	Date
Port Glasgow East Yard	R. Duncan & Company	1915
Marine Engine building plant	David Rowan & Company	1917
Inch Yard	Dunlop Bremner	1919
Glen Yard	William Hamilton Shipbuilders	1920
Entire company	J. Dunlop Steel Stockholders	1920
Port Glasgow Yard	Murdoch & Murray	1922
Entire company	Rankine & Blackmore Engine builders	1923
Entire company	Ayrshire Dockyard, Irvine	1928
Entire company	Steel Company of Scotland	1934
Entire company	Fairfield of Govan	1935
Entire remainder of Company	Beardmore	1935–6

Source: Derived from content in Murphy, "Déjà Vu all over again: The Reluctant Rise and Protracted Fall of Scott Lithgow" (PhD thesis, University of Westminster).

When the re-armament effort appeared on the horizon two years later, Lithgow sold the Steel Company of Scotland to Colville's and made a profit of more than forty percent – almost £300,000.[77]

A similar pattern was repeated with Fairfield in Govan. In the middle of 1935, Fairfield was working on a contract for a liner with the Anchor Line, itself supported by Cunard, the famous liner operators. Part-way through the contract, Anchor defaulted on bills of £133,000, which forced Fairfield – already at the limit of its £200,000 Bank of England overdraft – to the brink of receivership.[78] On one level, this episode should highlight the fragility of private armaments manufacturers even in 1935, given that a default on payment of a single order could have catastrophic effects. This was not unlike numerous other struggling yards at the time, but the difference was the way in which Fairfield was sold. It should be recalled that the Bank of England had struck a deal with Lithgow for his NSS scheme to purchase Fairfield in the future should it be unable to repay its loan. What transpired was that Lithgow privately enquired and then purchased the

[77] P.L. Payne, *Colvilles and the Scottish Steel Industry* (Oxford, 1979), 209–213; Murphy, "Déjà Vu all over again," 63 and 68; and Reid, *James Lithgow*, 169.

[78] S. Diaper, "The Sterling Combine and the Shipbuilding Industry: Merchant Banking and Industrial Finance in the 1920s," in J. Van Helten and Y. Cassis (eds.), *Capitalism in a Mature Economy: Financial Institutions, Capital Exports and British Industry, 1870–1939* (London, 1990), 87.

yard from the Bank for himself (although Sir Alexander Kennedy was retained as the firm's chairman until 1937, when Lithgow paid him a sum of £5000 to leave), rather than on behalf of the NSS.[79] Purchasing the shares cost Lithgow £245,000.[80]

Like his reorganisation and mothballing of the Palmer yard in 1934, there appears to be no obvious outward justification for Lithgow taking over another yard in 1935. After all, it flew in the face of the NSS scheme to rationalise the industry by removing yards that were not financially viable – and Fairfield had proven for some years that it was not viable without Bank of England assistance. As a merchant or passenger shipbuilder, there remained more than ample capacity in surviving firms – the NSS continued to liquidate capacity of this kind until 1939.[81] In terms of warship and armament facilities, Fairfield was an altogether different proposition, but once more there are logical inconsistencies. The NSS claimed in 1933 that more needed to be done to reduce berths in the warship and liner sectors,[82] yet when he had the chance to do just that, Lithgow stepped in to preserve capacity instead. Moreover, Lithgow had no previous background in warship construction and had never (before 1933) shown any interest in acquiring any of the other struggling yards. As this pre-dated Abyssinia (and the re-election of the National Government), it also pre-dated the much stronger public commitment to rearmament in the Second White Paper. So, even in the second-half of 1935 there were no interested parties. This underlines that private industry did *not* expect the yard to profit from a large-scale rearmament soon – Lithgow was the only new entrant to the warshipbuilding sector in the years between Washington and the Second White Paper. Moreover, it turned out to be a very good investment indeed: total profit in the decade after Lithgow took over was in the region of £1.5m, including money set aside for depreciation and contingencies. In the five years before he took over, it was £90,000.[83] Therefore, the justification his biographer has subsequently given for the purchase, namely that he did it out of concern for the workers and economy of Govan,[84] seems highly implausible.

Of course, the evidence is circumstantial. There can be no direct link made between his work with the PSOC and his purchase of Fairfield, save that the remaining justifications seem far more inconsistent with Lithgow's previous behaviour than the remaining alternatives. In this sense, it is impossible to say whether retaining Kennedy as the firm's spokesman and figurehead was a deliberate calculation to learn more about how an armaments manufacturer

---

79 BOE, SMT3/282, Norman to Duncan, 18 July 1932.

80 GCC, UCS2/1, Fairfield, board minutes, November 1935. Lithgow took 24,580 shares at £10 each.

81 Slaven, *British Shipbuilding,* table 4.9.

82 NMM, NSS, minutes, 16 December 1932. For an example of the notice sent to the warshipbuilders, see GUAS, GD320/2/10, Lithgow to Scott, 23 January 1933.

83 Derived from Slaven, *British Shipbuilding,* 111.

84 Reid, *James Lithgow,* 171–172.

operated, was imposed upon him, or even was to deflect attention elsewhere. His purchase of Beardmore, however, strengthens the case that he at least utilised his inside information before 1936 to indirectly assist the PSOC and Supply Board by taking a personal involvement in the manufacture of naval armaments.

As already discussed, Lithgow was invited onto the board of Beardmore back in 1932 by Sir Andrew Duncan at the request of Montagu Norman. By 1935, the Bank of England (via the Securities Management Trust) had pumped several hundred thousand pounds – again at Duncan's suggestion – into the larger Beardmore company to avoid collapse, although the sale of the shipyard to the NSS had brought back £200,000 of that in 1930. However, the remnants of the company, including the steel forge in Parkhead, were still a financial black hole for the Bank. It was therefore desirable that instead of scrapping the plant, some form of rationalisation akin to what was being attempted in steel and through the NSS be applied to what remained of Beardmore, and thus Lithgow's experience and connections made him an ideal candidate for the Bank to install on the board. Lord Invernairn, the previous owner, had been moved on – but between 1932 and 1935 Lithgow, pre-occupied with the NSS and his other work, played only a minor role in company policy. During those years, the man charged by the Bank with rescuing Beardmore was a German by the name of H.A. Reincke, who served as chairman between 1930 and 1936.[85]

Like his sudden shift in thinking regarding the mothballing of Palmer and Hebburn, it appears that Lithgow sensed an opportunity to become more involved with Beardmore after 1935, and seized upon it. Between 1932 and 1934, Lithgow and Reincke appeared to have a perfectly normal working relationship, but this deteriorated quite suddenly in 1935.[86] While the reasons for the deterioration are not absolutely clear, Lithgow, by his own admission, criticised Reincke's running of the company to Norman and Duncan throughout the year, and by the end of it had refused to work under him any longer.[87] Lithgow retrospectively justified his cooperation before that point on Reincke being "the nominee of my friend, Sir Andrew Duncan, who had confidence in him." However, Lithgow became convinced that Reincke was not "persona grata with the Heads of Government Departments and other large clients" and thus he "formed an opinion on which [he had] consistently acted that any important negotiations had to be conducted by himself." Furthermore, Lithgow intimated that he wanted to appoint a managing director to handle the London operations of both Fairfield and Beardmore in a

---

[85] Hume and Moss, *Beardmore,* 215.

[86] *Ibid.,* 236.

[87] Lithgow's admission came in a personal memo penned in 1941 in advance of a court case brought by Reincke for wrongful dismissal. Lithgow's personal recollections survive, although the case was eventually settled out of court. The Beardmore archival collections have retained these confidential letters, as well as notable correspondence among Lithgow, Invernairn and the Bank of England that shed considerable light on the reasons behind the sale of Beardmore to Lithgow in early 1936.

"more limited manner" than the previous two separate post-holders, and he did not think Reincke was up to that job.[88] Thus, when Lithgow got the chance to acquire the majority of shares in Beardmore from the Bank of England, he did so immediately, and Reincke was unceremoniously told to resign the chairmanship by Montagu Norman so that Lithgow could take his place.[89]

Lithgow's actions, from his increasing involvement with the company to ousting Reincke, should be placed in the context of wider supply planning. As had been the case with Fairfield, Lithgow negotiated personally with Norman for the sale at the end of 1935 before the Second White Paper was published. Norman had taken large holdings in Beardmore only, in his own words, "reluctantly and under great pressure" to avert "a catastrophe to the West of Scotland."[90] Lithgow's offer to take Beardmore off his hands, as it were, was therefore readily received. In a letter to Norman, Lithgow stated that he was willing to "infuse an active reconstruction and business-seeking policy" which could only be achieved "under the guidance of one constantly on the spot, with both the incentive and steadying influence which such a financial stake in the concern ensures." This was the role Lithgow was willing to assume personally. To finance the deal, Lithgow stated that he was willing to put £700,000–800,000 into Beardmore, but only "on the condition that for my hard cash I get undoubted security." To this, he said he was "averse to purchasing stocks or shares of doubtful value even at what may be regarded as a low value today, since it may prove to be over-valued at a later date." He was thus only interested in First Mortgage Debenture Stock held by Norman and the Bank of England. Moreover, he demanded the Bank's interest certificates at a "sufficiently heavy discount" so he could afford to forego first claim on shares or obligations and sweeten the deal to other shareholders. This, he believed, would mean that he would be "in the position of being interested in the equity in return for my efforts and services rather than in return for hard cash which I might have put up for a doubtful asset."[91]

---

88 GUAS, GD320/2/10/6, memorandum of R. v. L., 26 May 1941.

89 GUAS, GD320/2/10/6, "Minutes of Special Meetings of Directors of Beardmore, held at the Bank of England," 26 February 1936. It should be noted that although Reincke was removed from the chair of Beardmore, he was not sacked immediately – he remained for three years in ever less powerful positions. Lithgow quickly moved him first to Beardmore's London office (initially equal to the Glasgow Office which was headed by D.M. Cameron). Cameron was then made company secretary, and an assistant was appointed for his role, while Glasgow was redesignated the firm's 'Head Office'. By 1939, Reincke's role in the restructured organisation was now below Cameron's assistant (a Mr T. Sinclair), and his job was essentially to act only on instructions from the assistant secretary in Glasgow to have conversations with lower-ranking admiralty staff. In short, it was a huge demotion. This is the background to Reincke's wrongful dismissal claim which began in 1940.

90 *Ibid.*

91 GUAS, GD320/2/10/6, Lithgow to Norman, 15 February 1936.

Norman was more than happy to agree to these terms, for they appeared to solve a dilemma the Bank had been fretting over for some time. Lithgow was thus given a discount of £6000 on the First Mortgage Debenture Stock and a £63,000 discount on the certificates held by the Bank of England. He was also provided with assistance to borrow up to £250,000 should the firm require it. All in all, Lithgow paid just over £785,000 for a very high degree of control of Beardmore[92] – certainly far more than he did for Fairfield, but in the context of later rearmament orders, still a small sum.

To try to placate those hostile to the takeover, Lithgow intended to invite the previous owner, Lord Invernairn, back onto the board. Invernairn retained some voting rights in the company, and his cooperation would thus have been a boost to Lithgow.[93] Invernairn, however, was furious with how the sale had been conducted. He wrote back to Lithgow claiming that it would never have happened had Lithgow not done so well out of Colville and the Steel Company of Scotland merger, and most importantly that he had, in fact, also attempted to buy Beardmore back.[94] He wrote that:

> ... when I [Invernairn] desired to purchase these very same debentures that were held by the Bank of England, the Governor [Norman] virtually refused to consider the matter. You can appreciate my feeling on hearing that he has been willing to deal with you and actually disposed of them to you.[95]

In other words, Invernairn accused Lithgow – with some basis in fact, it must be said – of getting preferential treatment to buy Beardmore through his prior connections with Montagu Norman. As such, he felt that he could only accept a position if some of the company's money was held by independent trustees who were jointly appointed.[96]

Lithgow denied Invernairn's version of events, and claimed he was only doing what was necessary to restore "some measure of prosperity to the industry on which so much of the neighbourhood depends."[97] Lithgow felt that to achieve this he would need some "elasticity in control" which he could not have if the trustees held financial interests in the company. Lithgow claimed that he took on the role of leading Beardmore because he "was moved by a sincere desire to assist the West of Scotland through helping to re-establish the prosperity of a

---

92 GUAS, GD320/2/10/6, Norman to Lithgow, 21 February 1936.

93 Hume and Moss, *Beardmore,* 238.

94 In the event it was perhaps better for Beardmore that Lithgow assumed control, for Invernairn died somewhat suddenly aged 79 just a few days after his last correspondence in April 1936.

95 GUAS, GD320/2/10/6, Invernairn to Lithgow, 4 April 1936.

96 *Ibid.*

97 GUAS, GD320/2/10/6, Lithgow to Invernairn, 7 April 1936.

concern whose name is a household word amongst us." He did, however, admit six weeks earlier that he could not expect that he would be widely credited with such "publicly spirited motives" so it was for that reason he took the decision to link his financial fortunes with the fortunes of Beardmore.[98]

Of course, it is also probable that the prosperity of the West of Scotland did indeed matter deeply to Lithgow, but perhaps not as much to put over one million pounds of his own money into two firms with no guarantee of future profits. Still, there is absolutely no doubt that Lithgow knew the potential value of Beardmore more than anyone else involved with the company, including Norman. During the discussions to purchase Beardmore, Lithgow sent Duncan material marked "secret" from the PSOC detailing Beardmore's potential role in the (then) deficiency programmes from 1935. The site investigation, carried out by the Director of Ordnance Factories for SC I, noted that the firm was "in such a state of fluidity" that they could not accurately say what its future capacity would be. Thus, the report suggested omitting Beardmore from capacity calculations for the time being "owing to the disposal of facilities either through financial stress or the sale of essential equipment."[99]

This was a shrewd move on Lithgow's part. He was essentially raising the spectre of Beardmore having no future value and missing out on orders, while simultaneously offering to purchase the company from the Bank of England and "save" it from Reincke. Lithgow thus told Duncan that he had intimated to Reincke back in July 1935 that:

> … they should remove from the minds of the Government Departments any suggestion that their financial position precluded them from entering into Government contracts, and that on the contrary they should find out what the Departments really required … and put up equipment to meet such cases. I emphasised that I made that suggestion from inside knowledge of the developments which were taking place in the organisation of future armament supplies.[100]

Lithgow then told Duncan that the "chairman" (although this could mean chairman of SC I, SC III, the Supply Board or PSOC) told him two weeks earlier that they could not source bullet-proof plate from Beardmore.[101] This extraordinarily frank admission of what Lithgow knew is revealing for several reasons. First, it highlights the relative lack of influence he had in Beardmore until that point and explains why he was persuaded to overthrow Reincke in 1935. Second, it shows that Reincke was not taking his claims of "inside information"

98 GUAS, GD320/2/10/6, Lithgow to Invernairn, 26 February 1936.

99 GUAS, GD320/2/10/6, Lithgow to Duncan, 7 February 1936.

100 *Ibid.*

101 *Ibid.*

seriously. Third, that Lithgow did indeed share a small amount of his PSOC work with close business colleagues friends, and conversely, that Duncan knew little of armament supply organisation in 1936. Finally, this showed that Lithgow was acutely aware of the value that Beardmore held and would pay a significant sum to take control of it.

Before Rearmament: Constraints, Indecision and Planning without a Plan

Lithgow's accession to the helm of Beardmore occurred just two weeks before the Second White Paper. While this was contemporaneous with the debate on the future organisation of the CID (see below), it offers a convenient point for a short retrospective on what may best be termed as the "DRC era" after the three reports which helped shape rearmament as a policy and which punctuated the years 1933–1936. This was the period of "planning without spending" when industrialists operated in tandem with civil servants and service personnel within the CID framework to think through the major problems, but without the tools to directly solve them.

This is not to say that a great deal of supply and defence planning was not going on behind the scenes between 1932 and 1935 – the DRC, CoS and PSOC all played important roles in establishing the nature of British defensive deficiencies, including the industrial constraints on the speed of any future rearmament programme. The problem was not with any of these bodies, with the Advisory Panel of Industrialists, or with the Supply Committees that operated beneath the PSOC. These groups, flaws and minor disagreements aside, planned as effectively as their remit allowed and successfully identified issues, highlighted bottlenecks, and sought out sources of supply. The problem was with Cabinet, which frequently listened to and agreed with the various conclusions, but did not until late 1935 act upon the recommendations.

Not many can escape blame here: Chamberlain and the Treasury rejected the findings of DRC reports or sought to alter the conclusions to lower costs throughout of most of 1934 and 1935; the Foreign Secretary from 1931 to 1935, Sir John Simon, was so fixated with enticing Germany back to the disarmament conference that he railed against any re-equipment programmes that might antagonise Hitler;[102] and both Prime Ministers MacDonald and Baldwin at various points fretted about the National Government's record at the polls and commitment to the various disarmament pacts.[103] These are not exhaustive examples, but they do highlight

[102] D. Dutton, "Simon and Eden at the Foreign Office, 1931–1935," *Review of International Studies,* 20/1 (1994), 43.

[103] Robertson, "The British General Election of 1935," 174.

factors that prevented a coherent policy to address industrial weaknesses which would eventually hinder the rearmament effort.

However, although the DRC attempted to take a holistic and rounded view of deficiencies through investigations into capacity, facilities and skills, the Admiralty probably received a better share of the increased programmes than the naval threats at the time dictated. Credit must go to Maurice Hankey, who continually fought in the Navy's corner, and it should be remembered that at a time when air power and aerial deterrents were the most publicly acceptable face of armaments, it was the Royal Navy that received the largest share of the budget. One should not overstate progress here, though; for industry, even these successes fell short of what was necessary to rectify deficiencies in key bottlenecks.

From a business perspective, it is far from obvious that private armament manufacturers believed massive rearmament was a possibility before March 1936: Beardmore's chairman failed to listen to genuine inside knowledge in July 1935, Fairfield's directors and shareholders sold out just months before major orders resumed, and aside from Invernairn trying to buy back his former company, there was little interest from anyone except Lithgow in taking on more shipbuilding capacity. This must underline the continuing uncertainty that existed in the sector, the hand-to-mouth nature of business prior to rearmament, and the failings of government to address problems which had been identified several years earlier.

In this regard the relationship between the Admiralty and industry is conspicuous by its absence. Not since the destroyer rota ended some six years earlier had the two collaborated meaningfully. Neither side had the means to so. It is also clear that the Bank of England's senior directors, including Duncan, had only a very limited knowledge of rearmament plans before the Second White Paper, to the extent they were afforded only small glimpses of the organisation through snippets in Lithgow's letters. This is testament to the secrecy (Lithgow's leaks aside) of the CID machinery. In short, it says a great deal about the nature of the military-industrial relationship: for most parties, Weir, Lithgow and Balfour aside, there was none to speak of.

Finally, this period also represents another stage in the growing influence of civilian industrialists. If the Advisory Panel's first report seemed inconceivable during the Labour Government of 1929–1931, then Lord Weir's accession to the DPR in 1935 was a new level entirely. By contrast, the private naval armaments industry, even in 1936, still existed outside of the corridors of power. The crucial point, however, is this: Lithgow's acquisitions changed the nature of the game, for they offered the first genuine points of crossover between "outsiders" and "insiders." Moreover, they occurred just as rearmament was finally getting into gear.

Part Five

The Inskip Era and War, 1936–1941

Chapter Nine

The Minister for Coordination of Defence and Early Rearmament, 1936–1937

To be generous to the National Government's track record on defence policy between 1931 and 1935, the desire to keep publicity of deficiency and rearmament plans to a minimum was not wholly irrational. The issue was that the political and economic case for limited rearmament progressively strengthened from mid-1934, and not doing so began, by 1935, to badly hamper the development of an industrial skills and facilities base. Nevertheless, it was not entirely negative. A more welcome side effect of "planning without spending" was the involvement of Weir, Lithgow and Balfour with the CID, although this advantage could arguably have been better used by implementing policies like educational orders at an earlier stage. Of course, Lithgow used his involvement to take calculated risks for himself, so the net result was still the preservation of at least some industrial capacity that might otherwise have been lost.

If one can provide reasonable excuses for government inaction until 1935, it becomes harder from 1936. That year should, in theory, have been when planning and action finally came together, when much red tape was cut, when industrial deficiencies were properly rectified, and when rearmament began in earnest. Yet the National Government – following the Second White Paper's publication – still took several months to get a firm grip on rearmament policy and made some inexplicable mistakes that negated much of the positive work in earlier years.

Appointment of Minister for Coordination of Defence and DPR Developments

The National Government's first major step following the March announcement of rearmament was to appoint a new minister to oversee the increasingly complex and gargantuan CID organisation. Sir Thomas Inskip was installed as the newly-created Minister for Coordination of Defence" (MFCD) ten days after the White Paper was announced in Parliament. In principle, this was a sensible move, although it was at least a year too late. Moreover, just as the content of the Second White Paper had only taken shape in the weeks preceding publication, the decision that a new minister was necessary was similarly hurried. While this was a welcome dose of action from the government, from the perspective of defence "coordination" it would have been more advantageous to have appointed a minister at the time of the first DRC report (or at least after the First White Paper) when the competing hypotheses of defensive priorities first came to the fore. Instead, while the Second White Paper had outlined the rationale for a minister to oversee coordination, the genesis of the idea emerged out of a series of debates in the media, the Commons and in private just a few weeks prior to the new minister's eventual unveiling.[1]

The wider political context here was that Baldwin and the government had been savaged by press and public since the end of 1935 for appearing militarily weak in the face of the Abyssinian crisis and the subsequent disastrous attempts to bring an end to the hostilities.[2] At around the same time (January 1936), there was an outcry – led by former Chief of Air Staff Lord Trenchard – that defence coordination was unsatisfactory and that the air force was being "starved" of funds. Trenchard cited both indirectly and directly the failings of the CoS and Hankey in supply organisation.[3] Though once a very senior RAF figure, Trenchard had retired well before Manchuria and thus had little direct experience of developments since 1932, although he had in the early 1920s clashed with Hankey on CID matters. Indeed, his most recent public intervention was in the Manchurian debate, where he argued for aerial rather than naval defence of Singapore and further angered the Cabinet Secretary with his involvement.[4] The Chancellor, Neville Chamberlain, was not unsympathetic to Trenchard's views on the primacy of air power and harboured his own desire for some reform of the CID and a return to focusing on cheaper aerial defences at home rather than a three-pronged rearmament across land, sea and air.[5] Biases aside, calls for an enquiry and overhaul of the CID's structure had been loud enough to reach the House of Commons by mid-February, and it was at this point, three weeks before the publication of the White Paper, that the proposals for a new minister took shape.

In the Commons, the criticism was specifically that the CoS were reaching agreements that were the "aggregates of the three service plans," leading to less than ideal compromises instead of the best possible decisions.[6] Both Neville Chamberlain and his brother Austen were vocally critical of the situation, the

---

1 S. Greenwood, "'Caligula's Horse' Revisited: Sir Thomas Inskip as Minister for Coordination of Defence, 1936–1939," *Journal of Strategic Studies,* 17/2 (1994), 22.

2 In particular, the Hoare-Laval pact, Foreign Secretary Samuel Hoare's secret agreement over Abyssinia with the French PM, Pierre Laval, in December 1935.

3 Roskill, *Hankey, Man of Secrets, 1931–1963* (London, 1974), 202.

4 J. Naylor, *A Man and an Institution: Sir Maurice Hankey, the Cabinet Secretariat and the Custody of Cabinet Secrecy* (Cambridge, 1984), 130.

5 Greenwood, "Caligula's Horse," 20.

6 *Ibid.*, 20–21.

latter attacking Baldwin's failure to oversee the organisation effectively. In the face of such criticism Baldwin was forced to admit that the other pressures of his role meant he had not given as much time to the affairs of the CID as he would have liked. To satisfy his critics, Baldwin suggested that there might be a minister who could concentrate wholly on the CID and report upwards to him to get the "best out of the committee."[7]

In the event, the characterisation of the CID for seeking compromise was not far from the truth, although it was certainly nothing new; the CID's subcommittees had been compromising over competing demands for years before 1935, even when the compromises, like the demands, were merely hypothetical. Moreover, having another minister in Cabinet to be another mouthpiece for the CID may even have been desirable, at least if it helped push the agenda for overcoming anticipated supply bottlenecks. This aside, criticism from both Chamberlains was misplaced and unfair. Given the competition for the same pool of finite resources, one could speculate as to the potential state of British defences had the CID's supply committees been completely bogged down in inter-departmental rivalry rather than reaching mutually satisfactory outcomes. Indeed, in terms of greasing the wheels of supply organisation, the ability to compromise with relatively few stalemates, was, with hindsight, laudable. The Cabinet and Neville Chamberlain's own Treasury could certainly have learned something from the ability of the CoS and the PSOC ability to reach agreements, for it was the indecision and lack of agreed policy at the top level that was holding up the work of the CID and its subordinate bodies, not the other way around.

Not surprisingly, though, Hankey and the CoS resisted any external attempt to meddle with the CID structure and sought to limit the power of any incoming minister over CID affairs. Hankey wanted to avoid the distraction of external review and the more serious overhaul that Chamberlain desired, and thus offered two options: that the CID accept a minister who would act as a deputy to the PM and chair the meetings of the DPR in the absence of Baldwin, or to accept Hankey's resignation. This offer allowed Baldwin and the Cabinet to head off outside criticism over the CID's decision-making process without, in Hankey's words, "upsetting the psychology of the whole machine." It also avoided a bitter dispute with Hankey and the CoS.[8] Somewhat ironically, then, the organisation that supposedly compromised too much managed to get Baldwin and the Cabinet to agree to a compromise for the new minister and kept the mechanism for formulating defence hypotheses and organising supply largely unchanged.

Matters then progressed quickly. Within four weeks of these discussions, the White Paper had been published and Sir Thomas Inskip was appointed. A lawyer by training, Inskip had none of the background in defence that ordinarily would have marked him out for such a role, but he was – again ironically – known for

---

7 TNA, CAB23/83/7, CC 7(36), Cabinet Conclusions," 17 February 1936.

8 TNA, CAB21/424, Hankey to Fisher, 14 February 1936.

his conciliatory approach to disputes and willingness to take advice.[9] Moreover, he had no inherent connections with or biases towards any of the Service Departments, and, as the official historian of rearmament put it, he was never likely to interfere in the normal workings of the committee.[10] He has, as a result, also been characterised as "pedestrian" and "uninspiring" in his role, but he could equally be viewed as a man unlikely to upset the applecart or increase the Prime Minister's workload at an already critical time. If Austen Chamberlain wanted a man of action to make the right decisions rather than just those mutually agreed upon, then Inskip was from the outset the polar opposite. He was nevertheless in a position of considerable power in terms of the day-to-day running of important subcommittees, and the CoS, Hankey, and Lord Weir would all have to work closely with him.

In sum, it is hard to escape the conclusion that Inskip's appointment was a knee-jerk reaction, that he was unlikely to be the man to drastically speed up the rearmament process, and that if Baldwin was fully committed to the best possible coordination of supply and defence policy in February 1936, the decision would have been handled differently. Moreover, Hankey must receive a portion of the criticism. What the rearmament drive needed was action, yet Hankey perceived the new minister as a threat to his organisation and resisted change, missing an opportunity to push the CID's agenda in Cabinet.

DPR, Advisory Board Developments and Lithgow's Business Interests, 1936

With rearmament now in the public sphere and with a minister, at least notionally, appointed to head the rearmament drive, 1936 was comparatively uneventful in terms of the PSOC and the supply subcommittees. While the previous years had been characterised by much planning in secret based on loose and ever-shifting hypotheses but very little spending, the task from April 1936 onwards was much clearer. It centred upon how to implement the recommendations of the third DRC report and the content of the Second White Paper. In naval terms, this meant building the DRC's "standard fleet," although the new battleships could not be laid down until the expiration of the London and Washington treaty limits at the end of December.

In industrial terms, the existence of treaty limits enforced a period of waiting. This meant there was relatively little new construction of larger vessels in 1936, except for five cruisers already being built to the previous *Town*-class design under the treaty. A further fifteen *Tribal*-class destroyers – ordered one week after the publication of the Second White Paper – helped pick up the

9 Greenwood, "Caligula's Horse," 19.

10 Gibbs, *Grand Strategy*, 276–277.

slack.[11] Most notable, however, was that of these twenty ships, just one was placed with the Royal Dockyards. The other nineteen went to members of the WSBC, providing the single greatest boost to private industry since Washington. In fairness, this was not entirely a political decision, since the remaining Royal Dockyards were engaged in modernisation work on old ships, but it still was without precedent in the period since 1922.[12] Moreover, the differences between men like Weir and Lithgow, who knew about the capacity bottlenecks and the likely scale of future rearmament, and the rest of private industry had to some degree been diminished after 1936 because some future requirements were now public knowledge. Consequently, as the private naval arms industry was now able to prepare more securely based on a steady flow of orders in the medium term, the incentive for further WSBC collaboration remained, for in 1936 it remained unclear how long any revival would last.

Of course, Weir, Lithgow and Balfour were still advisers to and members of the PSOC, and by the summer of 1936 Weir played a prominent role in the DPR. His recommendations in January had been turned directly into content in the Second White Paper, and his task in the intervening months had been to consider further the question of industrial organisation based on a report he circulated the same month.[13] To this end, he produced further reports in April and May and contributed fully to discussions. The National Government's guiding principle for rearmament generally was that the "normal" business activity of the country should not be severely disrupted where possible and that rearmament should not impair the nation's economic health.[14] In this regard, Weir expressed his concern that the rearmament programme could not be completed in the five-year time frame and that the "most serious bottleneck was the shortage of skilled labour" which would be exacerbated by the demands on "articles of a precision character," such as naval guns and gun mountings. He submitted a brief resume to the DPR of the situation (see Table 9.1, below) for the industries affected by the White Paper to illustrate the scale of the labour shortages.

Weir went on to detail the barriers to solving the labour question, namely the issue of the dilution of individual tasks and the time lag between hiring apprentices and their becoming skilled employees. These problems were clearly not new, but little progress had been made since the individual Supply Committees began thinking about them in earnest well over a year earlier. Only once orders had been placed were apprentices taken on. In other words, despite years of planning, the private naval armaments industry was scarcely better prepared than it was immediately after Manchuria. The problem, simply put, was that more could and

11 Peebles, *Warshipbuilding,* 148.

12 *Ibid.,* 139.

13 CCC, WEIR17, DPR(DR)9, "Memorandum by Lord Weir on Industrial Production," 28 May 1936.

14 Gibbs, *Grand Strategy,* 301–302.

Table 9.1:
Lord Weir's assessment of industrial capacity, 1936

Section of industry	Current activity
'Bread and Butter' Steel	At record output and order books
Alloy Steel	Very busy
Forging and Stamping	Very busy
Machine Tools	Output limited by labour shortage
Electrical Engineering	Very busy
Automobile Industry	Running at record output
Airframe Industry	Full capacity and expanding on large scale
Aero-Engine Industry	Not yet fully stretched as a whole
Heavy general engineering	Still has a margin
Light general engineering	Busy
Textile Machinery	Still has a margin
Marine	Most of the best facilities well employed
Auxiliary Machinery	Generally full order books
Shipbuilding	Few empty slips, nearing limit for skilled labour
Railway Shops	Busy
Locomotive Shops	Still a large margin
Chemical Engineering	Fairly busy

Source: "Minutes of meeting held on 13 March, 1936," DPR(DR) 2nd meeting CCC: WEIR17, pp. 8–9.

should have been done *before orders were placed* to strengthen the industrial base on which future rearmament was built. Instead, the National Government chose not to interfere.

This was still not fully grasped at the DPR level. Although the decision to invite Weir into the DPR was certainly a positive step, for it put those with executive decision-making power together with someone who genuinely understood industry, its members did not yet fully understand the issues that the PSOC had been articulating for several years. The committee merely reiterated that it would be "very bad policy to buy finished munitions from abroad," while also expressing a reluctance to adopt the "semi-war" conditions that Weir believed would be necessary to organise industry most effectively. Most telling, however, was the rather basic nature of Weir's explanations, which indicate an almost complete lack of supply understanding among his fellow members. Baldwin asked if it was possible to calculate the productive capacity of the country, for "the committee had little knowledge" of it, while Weir explained to others – in response to Ramsay MacDonald – that simply finding more skilled men was not a routine task, and that "these questions could well be dealt with by existing organisations who were working well and whose procedure should not be disturbed."[15]

As Weir suggested, these were not topics that Supply Committee members would have struggled with in the spring of 1936. His outline of the remaining capacity in industry was nearly identical to discussions in the Supply Board some months before, so it is possible or even probable that the DPR could have found out the same information had its members looked for it – instead they relied on Weir to furnish them with basic information that the PSOC already had. In terms of optical glass (used for Admiralty rangefinders and army and RAF gun sights), the DPR, almost unbelievably, debated in late March and April 1936 whether it would be a good idea for Inskip to investigate capacity in private industry, since it was likely that "large expansion of production was a necessity."[16] This was several years after the same question had been considered, and firms investigated, by the individual Supply Committees.

This was not an isolated example. Later, Weir reported that he was "preparing a paper based on his discussions with the PSOC ... which would be ready for circulation at the next meeting." By May, he still had to point out that "there was no effective reserve of fully trained men that the services could draw upon."[17] Thus, rather than tapping immediately into the wealth of information that the PSOC had collected, the DPR spent a significant part of the first year of its existence feeling its way into problems which had already been identified elsewhere. In fairness, this was not a permanent failing of the new top-level committee – things certainly improved from 1937 – and in some cases the information the Supply Committees had collected with regards to capacity had been overtaken by subsequent events and were now out of date. Moreover, the SCs had first raised the labour issue in 1929 but took until 1935 to properly examine and understand it.

This cannot excuse everything. The fundamental point is that examples of the DPR working effectively to gain information from the subordinate bodies of the CID and then to *act upon them* are largely conspicuous by their absence throughout 1936 – so as far as naval manufacture is concerned, it was another opportunity missed and more time wasted to prepare the industry for the rearmament drive ahead. Weir was undoubtedly central in informing the DPR members of the nature of industrial problems, but whether they needed him to play this role and could not have given themselves a primer from existing information is another matter.

---

15 CCC, WEIR17, DPR(DR), 2nd meeting of the DPR, 13 March 1936, quotes from 8–9.

16 CCC, WEIR17, ID/G/167, "Record of meeting held," 26 March 1936.

17 Both quotes are from CCC, WEIR17, 20th meeting of the DPR, 7 May 1936.

The appointment of a minister, the formation of the DPR and the appointment of Weir to it changed the dynamic of the Supply Board and therefore the role of the Advisory Panel. Since late 1935, it had been Weir who was first asked to produce reports that would be read at the top-level of the CID, while two years earlier a similar request would have been handled by Weir, Lithgow and Balfour (Lord Riverdale) in conjunction with the Supply Board or the PSOC. Thus, rather than a bottom-up approach, whereby the committees handling complex questions turned to industrialists for advice and subsequently took those recommendations upwards for ministerial approval, the formation of the DPR at best occasionally streamlined tasks but at worst lost time and momentum while duplicating processes already in place. Unlike the PSOC however, the DPR had real power to act quickly, but this advantage could have been better utilised had those with relevant experience been given power earlier rather than simply forming a new top-level committee and essentially relying on Weir to be the PSOC's spokesperson.

Weir was emerging as Britain's pre-eminent industrial authority within the government. The power the DPR had to make decisions, coupled with his role as a full voting member with undoubted expertise, was a great advantage compared with his earlier advisory roles. For instance, when he suggested names of firms and individuals to chair a panel to help organise the steel industry, his recommendations were immediately implemented in full. Interestingly, his choice of chair for that panel was Sir Andrew Duncan, a long-term collaborator with Weir and Lithgow.[18] This was the kind of authority that the Supply Committees – and indeed his colleagues on the Advisory Panel – had sorely lacked in preceding years.

As an interesting footnote to the emerging supply organisation, there was one member of the framework who never believed Duncan and Weir to be quite as distinguished and deserving of praise as a great number of CID members clearly believed them to be. Sir George Turner, assistant to Sir Harold Brown (Engineer-in-Chief of the Royal Navy and chair of SC III), kept diaries and notes, initially to allow him to write his memoirs.[19] These offer a rare glimpse into the personal remarks and recollections which were often omitted from correspondence. Notably, he believed Weir to be a

> Peculiar influence ... he is treated like a master by those who attend the DPR committee. He sends for them. When he says turn they all turn.

[18] CCC, WEIR17, DPR(DR)9, "Preliminary List of Sections of Industry with which Early Contact should be Made," 28 May 1936, 13.

[19] Turner's diaries are in the Turner Brown papers at King's College London. They pertain almost exclusively to the war, but there are several pages not in diary form relating to the period after October 1936 which appear to be recollections. As the concluding section shows, these were almost certainly clouded by his later experience of the Ministry of Defence and the war, but Turner had rather less respect for Lord Weir, Sir James Lithgow and Sir Andrew Duncan than did Brown.

> Why? I expressed the opinion that he didn't know much about how to build an ammonia plant. The answer [from Brown] was "he is a director of Imperial Chemical Industries". His sole contribution on the ammonia question has been to press for ICI to be given the contract regardless of the existence of a lower offer.[20]

In Turner's eyes, it was "doubtful whether Weir's influence was as positive as his friends tried to make people believe." He was not just an advisor but an "interferer." The "danger" with Weir was that "his reputation and his manner persuaded most people that he could never be talking arrant nonsense so it was not recognised as such at the time." Perhaps not surprisingly, Turner believed Weir's influence over Sir Andrew Duncan to be enormous and that Weir had an acute weakness for "jobbery" – that is, selecting his friends for jobs "ten times too big for them." Duncan received similar treatment and was described as "mesmerised by Weir" and someone who "tried to be his own managing director" by treating his government work "like big business."[21]

Although Turner admitted that Weir clearly had experience, his damning assessment – of which more followed during the war itself – offers another side to the development of industrial influence within the state. In Turner's eyes, these "experts" were not subject to the same scrutiny as others, precisely *because* they were presumed to be authorities in their respective fields. It is not difficult to imagine – given the lack of expertise in the DPR – that Weir held the greatest influence at least partly by his supposed status. It has been argued here that industrial advisers were a very useful part of the CID organisation throughout disarmament and rearmament. But as Turner's assessment illustrates, this does not mean that civilian businessmen lost interest in business. They sought out allies, diversified their interests when it suited them to do so, found ways to solve problems that benefitted their own business interests, and were not afraid to use their status to work the market for government contracts to their own ends. Duncan, Weir and Lithgow undoubtedly helped preserve supply capacity and organise new streams for the benefit of Britain, but, like Lithgow with the NSS and Fairfield or Beardmore, they did not miss the opportunity to consolidate their own position – financially or politically – when the chance presented itself.

The Supply Committees spent a significant part of 1936 and afterwards working on several important topics, but the linkages with the DPR, notionally the top of the organisational pyramid, were not well developed.[22] The problem

20 BLH, TB, box 7, "Other Items not in Diary Form."

21 *Ibid.*

22 See, for example, TNA, CAB60/32, PSO(SB), 50th meeting of the Supply Board," 22 July 1936; and PSO(SB)596, "Ninth Annual Report of Supply Committee III," for examples of ongoing work.

remained the same: those at the cutting edge of investigations had no power, and those with power were not at the cutting edge, with Weir as the notable exception. This was not lost on those in the Supply Committees. In October 1936, Sir Harold Brown left his post and moved to a position in the War Office – despite having a naval background – and soon stopped consulting with the advisory committee structure altogether. Interestingly, the man behind the move appears once more to have been Lord Weir, who personally recommended Brown to the DPR.[23]

In contrast to Weir, Lithgow was becoming markedly less interested in CID affairs and far more interested in running his own firms (principally Lithgows, Beardmore and Fairfield). Weir, in a move indicative not only of his own role but also of the lack of advanced planning before Inskip's appointment, had first written to Balfour and Lithgow two weeks in advance of the Second White Paper's publication to warn of "a fundamental change in the organisation of which [the Advisory Panel] was an appendage," and suggested a meeting to discuss it.[24] He was referring of course to the debate about whether to appoint a minister to oversee, and perhaps even overhaul, the CID. Thus, Weir – through his DPR connections – was now rather closer to Cabinet thinking than either of his counterparts. It also underlines the lack of long-term planning in the Inskip appointment and Weir's feeling of compulsion to keep his fellow advisers abreast of developments.

The changing situation within the CID prompted Lithgow to confess to Weir that he now felt the whole committee structure was a "farce" and considered resigning.[25] In truth, only part of Lithgow's chagrin was down to his marginalised position on the Advisory Panel. He was also attempting to demonstrate – through his CID connections – that the vacant Beardmore plant at Dalmuir (constituting the dismantled shipyard that the NSS owned and the remaining marine engineering plant) would be an ideal location for expanding armaments production. He was particularly keen on supplementing the Woolwich Arsenal's munitions production, as well as securing orders for diesel engines and tanks. This was, of course, on the proviso that there was some state assistance with re-equipping costs.[26] Knowing his friend's influence, he had written to Weir outlining his case several times in the weeks after taking full control of Beardmore, and bolstered his plans with a site survey provided by the NSS's valuer, John Barr. Knowing also that the major Admiralty programmes were perhaps a year away, Lithgow was undoubtedly keen to find out what the other service departments needed in the long term so any re-equipping of the idle yard

23 Reader, *The Weir Group,* 129.

24 GUAS, DC96/21/9, Lithgow to Weir (citing Weir's earlier letter – not preserved), 22 February 1936.

25 *Ibid.*

26 GUAS, DC96/21/9, Lithgow to Weir, 2 March, 6 March, 11 March and 26 March 1936.

could begin as soon as possible. On this matter, Lithgow was finding it difficult to get firm answers; Inskip and the service departments were non-committal, and Weir could say no more than "from all he could learn," the naval "gun programme would keep Beardmore very busy." Weir's best advice was merely that Lithgow should "keep in closest touch with the [Master General of the Ordnance] and the Admiralty."[27]

Lithgow's frustration was not without foundation. Rearmament of a limited sort had begun in spring and summer, but comparatively little was achieved in the six months after the publication of the White Paper. It was late 1936 before the RAF's rearmament gathered any momentum – "Scheme F," the programme under which air rearmament was known until 1938, was only fully conceptualised in February 1936 after the new Air Secretary, Lord Swinton, was appointed. By that summer, the Air Minister was already considering whether to abandon the scheme and focus on developing new types of aircraft instead. The Army fared little better; political will for sending a force abroad was fragile, and within a few months of the White Paper the whole question of whether the Army should be ordering large quantities of tanks and artillery based on the principle of a continental commitment was re-opened, which naturally stalled all progress.[28]

Lithgow eventually lost patience in the autumn months of 1936. He wrote first to Weir, and then to Inskip, to vent his anger. After one meeting with Inskip at Weir's house in November, Lithgow formed the opinion that there was "no real long view planning being effectively considered" and that he could not begin training staff because it was "perfectly obvious that little or nothing is done ... until every 'i' is dotted and 't' is stroked in the specifications for individual requirements." He reserved particular fury for the PSOC, complaining that he had not even read the latest annual report until after it was produced and circulated, and that while the report claimed that "contact had been maintained with the Advisory Panel of Industrialists," it was clearly the case that "You [Weir] have been taken fully into the confidence of the fighting services, while so far as I am concerned the contact referred to is just a farce." At this point, he once more threatened to resign.[29]

Attempting to douse the flames, Inskip responded to Lithgow noting the "debt which ministers owed [him] for giving so much help."[30] Lithgow was still clearly furious that Beardmore's Dalmuir site was not "receiving the consideration it ought to receive" and threatened that without assistance the directors of the company would decide to "dispose of the plant piecemeal."

27 GUAS, DC96/21/4, Weir to Lithgow, 27 March 1936.

28 Gibbs, *Grand Strategy*, 303; and Postan, *British War Production*, 16 and 28–29.

29 All quotations are from GUAS, DC96/21/4, Lithgow to Weir, 10 December 1936.

30 GUAS, DC96/21/4, Inskip to Lithgow, 9 November 1936. The correspondence between Lithgow and Inskip was forwarded by Inskip to Weir, which is why it appears in his papers rather than Lithgow's.

While Lithgow "appreciated the difficulties which considerations of speed placed upon [Inskip] in regard of the wider questions of policy," he still believed the plant would be "broken up against the national interests" if the government did not act.[31] Weir, seeking to avoid any further deterioration in relations, told Inskip that he would "do his best to practise the art of lion taming," and by the end of 1936 Lithgow and Inskip had resolved – if only in part – their differences.[32]

Lithgow did, however, claim to Inskip that he had "no financial interest in Beardmore apart from having the responsibility of being Chairman," for it was the Treasury that held the stocks and shares, and was therefore "much more directly interested" than he was in the future of the company.[33] This was a blatant misrepresentation, for while the NSS owned the Dalmuir shipyard, Lithgow was clearly conceptualising the sites as part of the same firm as far as rearmament orders were concerned.[34] It is therefore hard to find an alternate explanation other than that Lithgow was attempting to mislead Inskip into believing the Treasury's own interests were tied up with Dalmuir being re-equipped to receive future rearmament orders.

Furthermore, it was not the case that the Beardmore plant in Parkhead – which he controlled – was to be idle. Entirely separately from Lithgow's discussions with Inskip and Weir, the Admiralty had indicated in the summer that Beardmore should double its capacity for armour plate to 9000 tons (as part of an expansion of the 1934 scheme for the three major armour manufacturers to agree to preserve plant and facilities), and that the costs of re-equipping were to be met by the Treasury. In addition, the howitzer plant and steel foundries were re-opened, while work on the enlargement of the armour plant began in October, all at a cost to the Exchequer of £143,580 by the end of 1936.[35] All in all, by December Beardmore was doing significantly better than twelve months earlier, before Lithgow assumed control.

In sum, while the relationship between one expert advisor and the man hastily appointed to oversee the coordination of defence was frosty, the larger problem was that the beginning of rearmament had significantly muddied the waters of supply organisation. The problem in 1934 and 1935 was that the top echelons of the National Government did not pay enough attention to the recommendations of the PSOC and its subordinates. In 1936, the problem was that they concentrated too much on the DPR. Weir's influence had grown to make him one of the DPR's central pillars, but Inskip and the rest of the committee were hastily learning the ropes and asking questions which had in

31 GUAS, DC96/21/4, Lithgow to Inskip, 17 November 1936.

32 GUAS, DC96/21/4, Weir to Inskip, 17 December 1936.

33 GUAS, DC96/21/4, Lithgow to Inskip, 17 November 1936.

34 Hume and Moss, *Beardmore,* 238. Earlier in the letter to Inskip Lithgow even referred to "his colleagues on the Board."

35 Hume and Moss, *Beardmore,* 238.

most cases already been asked, all while the PSOC and Advisory Panel were being marginalised. Expert advice, Weir's aside, was poorly utilised between March and December 1936. Given the protracted length of time taken to reach the stage of the White Paper and the enormous planning effort undertaken since Manchuria, this is baffling.

The Royal Commission on the Private Manufacture of Arms

At this juncture, attention should be turned to the Royal Commission on the Private Manufacture of Armaments (RCPMA). In October 1936, it published its report on the nature of profits in the arms industries. On one hand, the RCMPA was therefore a highly significant body, since it was an investigation into the supposed "evils" of the armaments industry, which by its very nature was a threat to the WSBC or worse still, was capable of derailing or hampering defence and supply planning at a critical point of rearmament. It is therefore impossible to ignore. On the other hand, one must be careful of attaching too much importance to its conclusions. The RCMPA was first devised in late 1934 in response to what *The Times* then called "genuine [public] sentiment of revulsion" against profit in arms,[36] but by the time it was ready to report nearly two years later, it was in many ways a relic of a bygone era of public pacifism. In terms of private naval manufacture and supply planning, the RCPMA did not derail, or even particularly adversely affect, the internal workings of the CID or the cartel-like nature of the WSBC. But an overview of why the commission "failed" and why the National Government and private manufacture emerged unscathed from the episode is essential, for it reveals much about the resilience of certain personnel within the CID and the nature of the political mindset in Britain.

The RCMPA was the National Government's public acceptance of the weight of public sentiment towards disarmament in 1934. It should be viewed as another by-product of the constraints the government believed it faced, along with the rather limited First White Paper on Defence (1935) and the private concern Baldwin, MacDonald and others had over the results of the by-elections and the "peace ballot" of the time. It also followed on from a similar enquiry in the United States, which had indirectly implicated Vickers and ICI in the misbehaviour of American firms seeking to bribe foreign customers into placing orders during the Great War.[37] In this sense, the calls for an enquiry were fuelled by historical allegations of wrongdoing and not necessarily by the current practice of armament manufacturers during disarmament. Thus, the

36 Anderson, "British Rearmament," 12.

37 M.W. Coulter, *The Senate Munitions Inquiry of the 1930s: Beyond the Merchants of Death* (Hartford, CT, 1997), 3.

commission was an anachronism almost as soon as the first private hearings took place in May 1935, let alone almost eighteen months later when the final report was published.

The commissioners – lawyers, correspondents and a businessman – were not experts in defence policy but were intelligent, sensible, well-meaning and quite willing to listen to the concerns of the wider populace. Fundamentally however, there were two major flaws in its thinking. The first is that the RCMPA understood private manufacture through the lens of the Great War, where claims that private profit had a hand in the origins and duration of the conflict could be tested. The second is that the evidence "against" private manufacture was always anecdotal, patchy and indirect, and this fatally undermined the case for greater regulation. For instance, a group of church leaders signed a memorandum calling profits in arms "repugnant," stating that these "encouraged war" and advocating a transfer to "public authority," but they cited no specifics. Lord Cecil's opinion was that the "public wanted the industry in public hands," but he admitted that he could not "profess to offer direct evidence of the evils of the question." The Labour politician William Arnold-Foster admitted that his case was not based upon evidence of wrongdoing but rather on simple common sense. The Communist Harry Pollitt performed better, showing that Britain's private armament makers had sold weapons to both China and Japan (belligerents in the 1931 Manchurian crisis). Beyond this, however, none could pin these facts to the argument that private firms deliberately conspired to create conflict for their own gain.[38]

This was typical of the broader case against private manufacture for the entire period between 1919 and 1935. Proponents of this position argued that it was logical and intuitive that industry needed war to survive, and perhaps even did survive through the sale of weapons abroad, but they provided no solid evidence that private profit *led to* war. Instead, the best they could do was to prove that the reverse was true. Potentially far more serious, however, was the testimony of David Lloyd George, former Prime Minister, War Secretary and Minister of Munitions. Testifying in May 1936 after the publication of the Second White Paper, Lloyd George had experienced first-hand the organisation of private arms manufacturers during the Great War. His opinion was that private firms could not be relied upon to meet the country's needs in times of emergency and that they had failed to cope with demand during the last war led to the conclusion that only nationalisation could provide a platform to meet essential needs during a major conflict. This carried potentially much greater implications for industry.[39]

The problem was compounded by less-than-stellar performances in 1936 from the representatives of the private manufacturers themselves. Vickers' chairman, the seventy-five-year-old Herbert Lawrence, appeared almost dangerously unhinged

[38] Anderson, "British Rearmament," 14–17.

[39] *Ibid.*, 19.

when he railed against a "pacifist prejudice" and admitted that Vickers might well go out of business if disarmament and peace were to prevail indefinitely. Lawrence, of course, had little background in wartime supply – he was in his mid-sixties when he joined the firm after a career in finance in the 1920s – but in a position of considerable authority, he perhaps did more to make the case for the other side than they did themselves.

Similarly, Charles Craven, Vickers' managing director, struggled to explain a leaked letter which appeared to show the firm hiding royalty agreements from the Admiralty. It was at this point that Craven apparently chose to disclose to the committee (via an appendix to written evidence submitted) the existence of the WSBC and that it had, in fact, been regulating workloads and prices, although no explanation survives for why he did so.[40] This much is at least present in the surviving WSBC papers. From this, it is probable that Craven was fearful of the repercussions if such a revelation came out during heated questioning, and he instead attempted to supply an explanation for the motivations behind the WSBC's formation – which were given as preservation rather than profit. Astonishingly, however, there was no further reference by defendants, the commissioners, or in the final report to this startling admission. Moreover, the appendix was never published. It is possible that the commissioners simply did not realise, buried deep within the mountain of statistics and material presented, what Craven was saying, but there were certainly no repercussions, nor did the Admiralty seem to change its relationship with the WSBC. It is also possible, given that the memorandum only appears to exist in the WSBC's records, that it was never actually submitted as evidence to the RCPMA.[41] This said, Hankey, when preparing evidence in the CID, clearly *believed* that armament manufacturers were "operating as a ring,"[42] but this is one of the few times such a belief

40 Murphy, "Déjà Vu all over again," 16.

41 Murphy and Johnman argue that the WSBC was revealed to the Admiralty after the RCPMA and was regulated thereafter. But the evidence for regulation is unclear, centring on a 1940 memo from Captain Crease, who was a senior naval officer; they claim that this hints at Admiralty regulation. Yet Crease had worked for the WSBC since the 1920s and coordinated many aspects of its work. Moreover, the profit margins continued to increase throughout the years 1937–1940, and there is no other reference to the existence of the WSBC until the Admiralty began to investigate the accounting records of each firm during the war. It is therefore at least plausible that the WSBC was not understood by the Admiralty during the RCPMA, and it was only when the cost investigations began once war was declared that the Admiralty fully grasped the nature of the group. Indeed, the WSBC's records stop rather abruptly at this point, which suggests that once exposed, the Admiralty exercised the right to fully inspect the books of each firm, destroying the mechanism to drive up prices. See Johnman and Murphy, "Welding," 101; and Murphy; "Déjà Vu all over again," 16.

42 TNA, CAB16/124, "CID – Royal Commission on the Private Manufacture of and Trading in Arms – Record of Meeting at Whitehall, Secret," 21 November 1935, 5.

has been recorded by any CID or Admiralty member in writing and pre-dates E.C. Jubb's famous line from 1936 that the system of competitive tendering remained "one of the greatest safeguards against undue profits."[43]

Weir, Lithgow and Balfour were not called to give evidence. This is not surprising. After all, none of them were known armament manufacturers in the conventional sense, and the RCPMA was looking backwards to the Great War rather than to the contemporary circumstances of the 1930s. Lawrence and Craven aside, most of the evidence from the industry was not particularly contentious and consisted merely of denying and rebutting the allegations put forward. The strongest case for the continuation of the current arrangements came not from industry but from the government – or rather from the CID.

Maurice Hankey, showing the same skill he did while drafting DRC reports arguing for imperial (and thus naval) defence and resisting an overhaul of the CID, expertly fended off the commissioners. In private, he frankly admitted to colleagues that he was primarily interested in making sure the commissioners had "no chance of recommending [the] prohibition of private arms manufacture," and he formed a subcommittee within the CID to draft answers to questions and provide evidence for the defence.[44] Hankey called upon PSOC members, including Dowding, Bovenschen and Hugh Elles to provide evidence and for the most part made short work arguing that the system in place worked perfectly well and that only private – not nationalised – armament manufacture could provide the necessary expansion in time of emergency. In their view, it was the "strong export trade" that kept afloat a nexus of capacity that could be called upon in war. Of course, while fine in theory, this skirted the issue that the government had not been willing to subsidise firms for the export of arms, and as such, the British private manufacturers were losing out frequently to foreign bids in Italy, the United States, and elsewhere.[45] Nevertheless, the ability of Hankey and his team to present consistent evidence was in marked contrast to those seeking an end to private manufacture.

In sum, the "prosecution" was poor, but the input from industry – particularly Vickers – nearly shot the defence in the foot. They were saved by a combination of Hankey's cunning, and, it must be said, by the fluid political situation which completely altered the groundswell of public opinion and National Government thought between November 1934 and October 1936. For while the commissioners only made some modest recommendations to better monitor the current system for abuses, even this was rapidly rejected – as the Cabinet, quite deliberately, submitted the final report to Maurice Hankey (notionally the CID) for "approval."

[43] CUL, MS Templewood, part 9, file, 5, "Admiralty Contracts, Prevention of Profiteering," 17 July 1936.

[44] TNA, CAB16/124, "Royal Commission on the Private Manufacture of and trading in Arms – Meeting at Whitehall," 21 November 1935; and Roskill, *Hankey*, II, 248.

[45] TNA, CAB60/13, PSO 361, "Report by Subcommittee on Industrial Intelligence in Foreign Countries," 13 March 1933, 7.

Since approving recommendations that he had so vigorously fought against was somewhat akin to a turkey voting for Christmas, Hankey gladly buried the report, and most of the recommendations never saw the light of day again.[46]

This episode highlights increasing cohesiveness in the Cabinet and a hardening of the collective mindset. In late 1934, when the Commission was reluctantly agreed to, the National Government was constantly fearful of public backlash (while also citing the adverse cost of armaments). Two years later, it could confidently ignore the findings of a Royal Commission. Even the recommendations on a system of costing to limit private profits, which although bureaucratic may have been desirable to the Treasury, were not acted upon. One presumes this was because rearmament had now reached the stage where, as Weir put it, the "word of the man in supply must carry." This contrasts markedly with the Inskip episode. It is apparent that the Cabinet was no longer willing to allow an overhaul or disruption to the system of defence and supply organisation, and trusted in it to deliver on rearmament. It is hard to imagine with the disagreements, indecision and inaction over deficiency programmes between Manchuria and Abyssinia that the government would have so quickly and unanimously buried something of this magnitude two years earlier.

Rearmament Proper and Admiralty Programmes

The RCPMA's report and eventual burial spanned the end of 1936 and the beginning of 1937. It was at this point, following the 31 December expiration of the London and Washington treaties, that rearmament finally got underway for most of the private naval armament manufacturers. 1936 had been better than any previous year since Washington in terms of new orders – but not by a huge margin. Expenditure on the navy stood at nearly £65m – the first time it had been above £60m since 1922 – but was still only £10m higher than the average spending between 1923 and 1935, and some £40m lower than the average for the first three years of the 1920s. Although a larger proportion of this pot was finally being funnelled towards private manufacture and away from the Royal Dockyards, it was still small in relative terms, and inadequate for preparing private industry for the challenges ahead.

There were some developments towards the end of 1936, however, that are not captured by naval expenditure alone. WSBC members, in tandem with Treasury funds, were beginning to renew and re-equip their yards in anticipation of future work. The sums spent on a per-yard basis by 1936 were modest compared with the size of orders that would eventually be placed – although these rose further once rearmament took hold. Beardmore spent under £200,000

46 TNA, CAB16/124, "Meeting of the Inter-Departmental Committee to Consider the Report of the Royal Commission," 22 January 1937.

on re-equipment costs in 1936 (and would eventually spend nearly £800,000),[47] while Cammell Laird spent £80,000 along with Vickers on meeting its initial share of extra armour plate requirements. Vickers Armstrong's first programme of naval re-equipment was undertaken in 1935 at a cost of £100,000 – but this would reach £750,000 by 1938.[48] In this respect, 1936 was the year when the renewal of facilities and the training of some new staff began in earnest, if not the year where construction took off.

As the figures illustrate, this was by no means complete by the expiration of the naval treaties; Beardmore, John Brown and several other yards were still renewing and re-equipping as late as 1938. One must therefore conclude that despite the efforts of the PSOC, the naval yards were not in prime condition by the time the treaties expired, and at any rate were not yet thinking of rearmament on a grand scale. In other words, the National Government had patently failed to do enough to prepare the industry between the First White Paper and the expiration of the treaties some twenty-two months later.

The Treasury did not help with re-equipment costs purely out of concern for Britain's industrial capacity; it did so because yard owners ostensibly did not fully believe that rearmament would provide more than temporary breathing space. The experience of the Great War guided thinking. Lithgow, who knew the rearmament programmes better than most of his contemporaries, believed the total costs of re-equipping Beardmore by 1939 – estimated at over £600,000 – would be of little or no commercial value by 1940. He was thus unwilling to accept the offer of an interest-free loan – the Treasury's first offer – and demanded that the state meet the costs in full.[49] There was a similar situation at John Brown; Sir Allan Grant credited the Controller of the Navy, Sir Reginald Henderson, with "energy and foresight" for pushing through the terms for the expansion of the Atlas armour works, again paid for by the Treasury.[50] However, the WSBC had a remarkable degree of confidence in their rota, and at any rate the Second White Paper was taken to mean a prolonged rearmament effort, but it appears the Treasury was not in a position to argue the point if it meant delays. It was the latest case of the "word of the man in supply" carrying.

In January 1937, two of the five *King George V*-class battleships – the most advanced class of Royal Navy warship to that point, and, although unknown at the time, the last true class of British battleship – were laid down.[51] Unlike smaller (and cheaper) ships, which were covered under the WSBC rota, battleships were omitted. Given their size and cost, the placing of a battleship

47 Hume and Moss, *Beardmore*, 241.

48 Scott, *Vickers,* 222.

49 Hume and Moss, *Beardmore*, 241.

50 Grant, *Steel and Ships*, 85.

51 *KGV* was followed by the *Lion* class, all of which were cancelled as they could not be combat-ready before 1944. The world's last ever battleship, *Vanguard,* followed next, and was the only ship of its class.

order was a rare – and in many respects a political – decision. This changed the tendering process and as such did not map on well to the WSBC work-sharing agreement. For these reasons battleships were never seriously considered for inclusion in the rota. In any case, the class eventually comprised five ships, and since one of each which was placed in the five-remaining capital shipbuilding yards (Vickers Armstrong, Cammell Laird, John Brown, Swan Hunter and Fairfield),[52] it would not have made a great difference either way. Even this was not without controversy, for the size and cost of the ships – and the prestige attached to gaining an order – meant that those that missed out in January (three more were laid down in May, June and July 1937) were disappointed and angered by the decision. It should therefore be unsurprising to discover that a great deal of behind-the-scenes grumbling went on between this subgroup of WSBC members when the first two orders were placed on Tyneside and Birkenhead, as none were awarded to the Clyde yards (essentially John Brown or the Lithgow-controlled Fairfield).

The decision on the location and order of the battleship contracts appears to have been signed off by Inskip and was based primarily upon Cammell Laird's recent lack of large orders and John Brown's current contract to complete the passenger liner *Queen Mary*.[53] Stephen Piggot, director of John Brown's Clydebank yard, wrote to Lord Weir to express his disappointment and to press Weir on the possibility of expediting an order for a third ship.[54] Lithgow, who Charles Craven by then believed was "a very pushy man,"[55] wrote to Inskip to vent his anger at what he believed were grave inaccuracies in the decision-making process caused by the "intrusion of political interest" into Admiralty allocations. Lithgow claimed his frustration was not borne out of being a shipbuilder but out of his "previous endeavours to assist government policy of stimulating activity in the distressed industrial areas," and that he feared "political agitation" on the Clyde if the balance was not redressed.[56]

Lithgow's concerns, of course, played on existing DPR (and PSOC) fears over cooperation with the trade unions – something which Weir and Lithgow had earlier been keen to circumvent – and one must strongly suspect that Lithgow knew full well the buttons that needed to be pressed to sufficiently worry Inskip. Inskip's response made it clear that orders would soon be placed on the Clyde, and that if he had "transgressed in respect of the battleship that has gone to Birkenhead," he hoped Lithgow would "believe it was an error of judgment rather than by intention."[57] Quite apart from this being further evidence of Inskip's skill as a conciliator, it highlights a further change in the

52 McMurtie (ed.), *Jane's Fighting Ships*, 202.
53 GUAS, DC35/31, Lithgow to Inskip, 2 November 1936.
54 GUAS, DC96/21/9, Piggot to Weir, 20 October 1936.
55 Warren, *Steel, Ships and Men,* 248.
56 GUAS, DC35/31, Lithgow to Inskip, 2 November 1936.
57 GUAS, DC35/31, Inskip to Lithgow, 9 November 1936.

dynamic of the members of the Advisory Panel. Ostensibly equals in 1934, by the beginning of 1937 Weir was understood by private industry (including Lithgow) to be powerful and well connected, while Lithgow felt excluded and powerless. Weir was the one being complained to by the heads of WSBC yards; Lithgow was among those doing the complaining.

This was a brief episode of disquiet, for by the summer of 1937 a whole raft of orders for ships of all kinds had been placed in WSBC yards, most of which were now closing in on full capacity for the first time in well over fifteen years.[58] Illustrative of the nature of private shipbuilding was that demand for naval ships was supplemented by a rebound in global trade. The new orders for the merchant marine this created meant that prosperity returned as quickly as it had disappeared in 1922. Demand for material (especially armoured plate) and skilled labour – both in relatively short supply – rocketed, causing both wages and prices to rise, perhaps by as much as fifteen or twenty percent by the beginning of 1938.[59]

This was more than was passed on to the Admiralty, and it underlines the success of the WSBC in price-fixing. For while the price of a typical tramp steamer increased by around fifteen percent (broadly in line with cost and overhead increases) in 1937, the profit margins on Admiralty work were more than thirty percent, or roughly twice the level of cost increases.[60] Indeed, these levels were about three times higher than the typical margins on naval contracts before 1936, which themselves were still far better than the razor-thin profits on merchant vessels or the losses incurred on Admiralty contracts in the years before the WSBC. Moreover, although there was another, albeit brief, industrial recession in 1938, naval contracts held up remarkably well, and prices were not brought under control by the Admiralty until well into the war.

How and why this could continue lies as much with the politics of the CID and the National Government as it does with slack Admiralty practices and too much faith in the principle of competitive tendering.[61] Hankey remained staunchly opposed to outside meddling with the CID's structure, and once rearmament began there was a tendency within the CID (and, it must be said, the Treasury) to maintain the "status quo" as much as possible. Anything that unduly upset or hindered progress was deemed undesirable, and the National Government and Hankey both emphasised the importance of the full and willing cooperation of industry.[62] Moreover, the belief that firms required large economic incentives to expand production – as witnessed in the funding for new plant and facilities – carried much weight. As such, any unduly close

58 Peebles, *Warshipbuilding,* 148.

59 Derived from GUAS, UCS1/9/79, "Memorandum on Conditions Now Existing in the Shipbuilding Industry," Shipbuilding Conference Confidential Report, December 1938.

60 Peebles, *Warshipbuilding,* 140–142.

61 CUL, MS Templewood, part 9, file, 5, "Admiralty Contracts, Prevention of Profiteering," 17 July 1936.

62 Rollings, "Whitehall," 532.

scrutiny or control of ordinary business practices was rejected outright, while the government announced to the public that "to put difficulties in the way of industrial enterprise is to place on its shoulders a great responsibility."[63] In short, the lack of progress before 1936 and the worsening international situation thereafter had convinced policymakers that making more obstacles for the rearmament drive – which they believed still had to be delicately balanced between security needs and the "normal business activity" of the country – was to be avoided at all costs. It was the combination of excessive Admiralty trust in industry coupled with the pressures on the National Government that allowed the WSBC to continue for so long.

Perhaps just as notable was how underprepared Britain still was once rearmament had finally gotten under way. The predicted shortages of facilities, skills and materials were not alleviated by the years of planning by the PSOC and the Supply Committees. Bottlenecks existed everywhere. Guns, mounting and armour, items about which SC III had expressed grave concern in 1932 and 1933, were still the main issues more than four years later. Reginald Henderson told the Supply Board that it had reached the stage that "for the next three years [until the end of 1940], however many slipways were available, it would not be possible to complete the ships to be built under the naval hypothesis with their full complement of guns."[64] A few months earlier, Weir pointed out to the DPR that while progress in certain respects appeared satisfactory for *new* construction, when combined with the extra capacity needed for rearming older vessels with new equipment, the "magnitude of the task" was great. His calculation was that at least 450 naval guns and gun mountings would be required by 1940, which was "a very big job for the gun makers."[65]

Another aspect of the problem was that while the WSBC handled the prices and contracts for naval ships, members were still competing with the rest of the merchant sector to take as many orders as they could while the going was good. Thus, instead of Vickers handling exclusively naval work, adherence to the "normal" business cycle meant that they and others were also attempting to secure new passenger and merchant orders, stretching their available staff even thinner.[66] Lithgow, for instance, would not accept the Treasury's stipulation that if they met the cost of expanding Beardmore, the new plant could only be used for naval work. He demanded, and received, permission to use it for any orders Beardmore could secure.[67] Money was being thrown at the problem at the last minute, precisely because warnings had not been heeded earlier.

---

[63] Peden, *British Rearmament and the Treasury*, 83.

[64] TNA, CAB60/32, PSO(SB), 55th meeting of the Supply Board, 6 October 1937.

[65] TNA, CAB4/27, 36th meeting of the DPR, 18 March 1937.

[66] TNA, SUPP3/18, "Interim Report of the Shipbuilding Consultative Committee," 23 July 1937, 7.

[67] GUAS, UGD100/1/16, "Conditions for the Extension of Armour Plant," December 1936.

Although the attempt by naval yards to secure merchant and passenger work could be justified on one level as providing a steady stream of work between naval orders to keep staff employed, it was not without adverse effects. Crucially, the inefficient allocation of merchant and naval work meant some naval yards were working to full capacity while their merchant counterparts did not have enough work to justify training extra apprentices. Thus, the contract for the *King George V*-class had to be extended by almost six months past its original 1940 deadline, while an extension of three months was granted on top of the initial estimates for the completion of destroyer orders.[68] By 1938, a *Tribal*-class destroyer was taking as long to build as WWI-era super dreadnoughts, which were fifteen times larger.[69] Even when otherwise finished, ships were being laid up because no guns were available. In short, the Admiralty, and by extension the British taxpayer, was paying much more per ship but was waiting far longer to receive the finished product. Problems that were identified years earlier had not been satisfactorily solved.

These problems were not lost on Lithgow. He complained to Inskip that the issue was the uncertainty that was being created through "break clauses" to allow the buyer quick exit from a contract being written into orders. These were being justified as necessary under "government policy," but it did not give shipbuilders confidence to pass up other orders when they were presented to them. In Lithgow's eyes, it appeared that "Cabinet was still taking their cue from Lord Robert Cecil [a leading disarmament voice] and Company" and that the "Treasury were still restraining the services from a whole-hearted development of their programmes." Moreover, Lithgow believed that constant delays meant that "public opinion was ready for action" while the fighting services still had "scarcely a single approved design they could put in front of manufacturers." In short, he was chiefly concerned that "no concerted long-view plan existed."[70]

Lithgow, it must be said, was extremely adept at playing the "national interest" card. He made a deliberate effort to link the lack of organisation with the possibility of military catastrophe, public revolt, or both. He claimed he was speaking from a position of first-hand experience as a concerned citizen and that his overriding interest was not his own wealth but the security of the nation and the livelihood of his workers. Whatever his true motivations, it is hard to escape the conclusion that Lithgow was using the levers available to him to influence Inskip into giving firmer answers on contracts. In this sense, Lithgow's refusal to meet the costs for the expansion of Beardmore is the other side of the same coin; he was using his firms as bargaining chips by claiming uncertainty over future

---

[68] GUAS, UCS 1/1/2, "Minutes of Board Meeting at Brown, Clydebank," 17 February 1938.

[69] Peebles, *Warshipbuilding*, 141.

[70] Both quotes are from GUAS, DC35/31, Lithgow to Inskip, 17 November 1936.

defence plans, and attempting to use his prior role as an advisor to pressure an inexperienced minister into making assurances.

Regardless of Lithgow's manoeuvring, the fact that bottlenecks existed and mistakes were still being made in the finished product in 1937 was a failure of policy, not of industry. More specifically, it was a failure at the top-level of government. The PSOC, Supply Board and Supply Committees – including the civilian industrial experts – had all fulfilled their remits, investigated the problems of supply and suggested methods for its improvement. That these problems still existed can be attributed either to a failure to equip the supply committees with the tools (specifically, the financial tools) to build up rearmament capabilities in good time, and as such represented a lack of faith from above in the CID structure or a failing within the Cabinet to recognise the seriousness of the situation. By treating industrial preparedness and the political process of preparing the country for war as the same issue, it was not ready to rearm when the time came.

This can be seen clearly when Cabinet considered the 1937 construction scheme. When the First Lord unveiled his plans based on the DPR's recommendations to commence work on twelve large and more than thirty smaller vessels, Cabinet agreed that the Chancellor's request to lay down some ships later than planned so as not to "resort to emergency measures in relation to labour, or to undue diversion of shipbuilding from their normal channels," while citing the effect on merchant shipbuilding.[71] In other words, the first major rearmament programme still favoured business as usual for the manufacturers and made no explicit attempt to *expand* capacity in case of future need. As private industry had received precious little assistance during disarmament, it would be unrealistic to expect it to have built itself up before rearmament got underway without far stronger incentives than it was provided with. Since such incentives could only come from the National Government or from foreign naval orders (which were being lost to nations that provided subsidies to private builders, itself another of the PSOC's suggestions), one must conclude that the blame must rest with Cabinet and not the supply organisation framework in the CID.

[71] TNA, CAB23/87/10, CC 8(37), "Cabinet Conclusions," 17 February 1937.

CHAPTER TEN

Later Rearmament and War Supply Organisation, 1937–1941

The missed opportunities before and during 1936 were painfully apparent in 1937. Inskip was certainly conscious of the supply problems that arose, but he continued the theme of taking too long to do anything about them. The issues in the naval armaments sector clearly warranted special attention and did finally prompt some badly-needed action. But the impetus did not come from Inskip. It came from the unsung former chairman of SC III, Sir Harold Brown. A naval engineer and vice admiral, Brown is not a particularly well-known figure in histories of rearmament, but he oversaw a series of innovations in the second half of 1937 and into 1938 that better utilised industrial expertise and finally overcame the barriers to progress that had slowed the earlier work of the CID's supply planning framework. As such, the work of Brown and his colleagues cannot go unmentioned.

Brown, the Boilermakers and the Shipbuilding Consultative Committee, 1937–1938

Leading on from Lithgow's concerns over the future usage of Beardmore, Inskip was in increasingly regular contact with Brown in 1937. It will be recalled that Brown had served as the Chairman of SC III (Shipbuilding) until September 1936 and as the Director General of munitions production in the War Office. He also played a significant role in the discussions to form the Advisory Panel of Industrialists in 1933. In 1937, Brown was given latitude to work on a raft of proposals to turn over factories to armaments production, which included proposals for the restructuring of vacant space on the Clyde for gun mounting work for both the Admiralty and the War Office.[1]

[1] Brown, "Reminiscences," II, 122. Brown's recollections offer a very rare insight into the organisation and failings of supply machinery from someone who was involved from the outset.

Brown was keen to put the Beardmore plants to full use, but he complained of a series of obstacles in his way which, perhaps predictably, had been raised by his own committee at Supply Board level in previous years but had not been properly rectified. In the case of Beardmore, Brown noted that the vacant Dalmuir yard had been stripped of equipment before 1933 and now required a complete re-tooling, but he found the backlog on machine tool orders to be prohibitive. His experience brought him into full agreement with Lithgow, as he believed that there had been a general "lack of appreciation [at Cabinet level] that building up effective capacity needed a long-term programme." Brown thus sought closer cooperation with industry to achieve his supply goals.[2] Moreover, he was among a growing number of supply experts who recognised that industry could never be fully efficient unless the National Government allowed rearmament to take priority over the normal business activity of the country. Weir had already advocated such "semi-war" conditions in the summer of 1936, but the DPR had shot down the idea, and there had been little progress since.

Although now notionally working for the War Office, Brown was first and foremost a naval officer and understood the issues – and leading figures – in naval armaments manufacture better than most. He compared the Admiralty favourably with the War Office and believed that the existence of established armament industries and industrial experts meant the former could work to long-term plans while the latter was practically starting from scratch. Understanding that he would require any spare naval armament capacity for field artillery, and knowing full well the benefits industrial experts could bring, he helped form, under the aegis of his old SC III, a Shipbuilding Consultative Committee (SCC) to bring together naval – and for the first time, merchant – shipbuilders. They were joined by Admiralty and Board of Trade figures with the aim of working around longer-term problems and bottlenecks in shipbuilding supply – and thus potentially providing spare capacity for the War Office – to better organise for wartime conditions.

The SCC, which presented its first report at the end of July 1937, was comprised of five representatives from shipbuilding: Craven, Lithgow, Denny and the presidents of the Shipbuilding Conference and the SEF, Amos Ayre and Francis Pyman. They were supplemented by three Admiralty and three Board of Trade representatives, including J.S. Gillingham, who had taken over from Brown at the helm of SC III.[3] The SCC, like the Advisory Panel before it, was tasked with thinking about wartime conditions and given access to the reports and work of SC III on shipbuilding in preceding years.[4] The SCC's investigation found that while there were still plenty of available berths in British

---

2 *Ibid.*, II, 155.

3 TNA, SUPP3/18, "Interim Report of the Shipbuilding Consultative Committee," appendix I.

4 *Ibid.*, 5.

merchant yards, there was insufficient labour to staff them. For this reason, the SCC advised that reopening the merchant facilities closed under the NSS was not advisable, for "if a surplus of labour becomes available it would be preferable to increase the output of existing yards."[5] This was built upon by a later report in November that estimated that over 400,000 *skilled* staff in over 200 specialisations would be needed for shipbuilding and naval engineering.[6] Moreover, the SCC calculated that compared with normal business conditions, an extra 932,000 tons of capacity over and above the naval programme could be handled in non-naval yards, but only if the government allocated orders in the most efficient manner rather than allowing the "business as usual practice" of competitive tendering. Taken together, the SCC believed this would alleviate the significant bottlenecks and shortages that were already apparent in the merchant and naval sectors alike.[7]

In theory, this suited both the shipbuilders and the supply planners, for the yards would all be kept busy working on the type of vessel they understood best, the government would not have to revive defunct yards at great expense, and efficiency would be greatly improved. A second report, updating the estimates to factor in potential changes in the scale of future naval orders, was presented a year later in July 1938. This argued that anything up to 350,000 extra tons on top of the estimates could be produced within the first twelve months of an emergency if the industry was strengthened during peace. Again, however, the focus was a *war* scenario, not preparations before it, and it remained so. The second report noted that despite these warnings, the SCC "was not directly concerned with the condition of industry in peace time" and that it was "beyond their scope to do more than draw attention" to problems. The maxim that the normal business cycle of the country should not be interfered with if peace prevailed still held, and this left the proposals in 1937 and 1938 in limbo. Indeed, when the report was discussed by the SC III its members declared that since this was a matter of "high policy" the committee only had the power to send it to Inskip for consideration.[8] It should therefore be no surprise that Brown and Lithgow were growing increasingly disillusioned with defence planning.

Nevertheless, the SCC was at least a step towards thinking about a more efficient type of organisation, and it tackled a different set of questions from those which had been the focus of the original Advisory Panel. The industrial advice at Supply Committee level was still consistently well informed, innovative, imaginative and provided solutions which genuinely offered new ways to organise the provision of warships and merchant vessels. It was just not well utilised;

---

5 *Ibid.*, 7.

6 TNA, CAB60/45, PSO(SB)713, "Categories of Staff and Workpeople," 30 November 1937.

7 "Interim Report of the Shipbuilding Consultative Committee," 7.

8 TNA, SUPP3/18, "Second Interim Report of the SCC," 8 July 1938, quotes from 1–3.

even by the time of the SCC, when the system was under increasing strain, the reports were only compiled a year apart, in July 1937 and July 1938, respectively. Moreover, the constraints were not new; top-level adherence to economic and business orthodoxy still hindered progress.

Similarly, the timing of the formation of a new shipbuilding committee, some four years after the creation of the first Advisory Panel, highlights another important fact. Despite the scale of the problem, it took until well into rearmament for other industrial experts to play a substantial role in supply planning. This reiterates how rare the Advisory Panel was. During the "secret" deficiency phase, Weir, Lithgow and Balfour were unique, and "inside" knowledge of CID practices and defence hypotheses was largely kept within this trio. That major naval armament manufacturers – including the managing director of Vickers, no less – were not invited in until rearmament was embedded within public knowledge is testament to it.

Unsurprisingly, problems persisted. The DPR noted in September 1937 its concern at the "delays in gun production" at Beardmore and asked Weir to discuss the problem with Lithgow and to inspect the factory. Beardmore had been given twenty-five percent of the entire gun order (both naval and anti-aircraft), and it was believed the delays would push the programme back by at least six months. Weir's follow-up report noted that Lithgow was in the process of spending £2m on the plant and was now spending five days a week at Beardmore to rectify the problem; generally, the response "appeared to him to be very satisfactory."[9] These were recurring themes: planning had failed to overcome bottlenecks, while the members of the DPR were, in part, replicating the work of the lower orders rather than fully utilising the available expertise. Moreover, when all else failed they had to rely on Weir – now their most senior industrial figure – to have a frank discussion with his friend. It is hard to conclude that this was the most efficient or professional way to cooperate with industry.

It was because of the ongoing frustration at the lack of tangible outcomes that Harold Brown attempted to circumvent the CID system altogether. Brown valued the advice he was receiving but felt constrained by the inability to act as promptly as circumstances demanded. As an engineer, by his own admission he preferred working problems out rather than talking about them *ad infinitum* in committees. Late one night in 1937, with the Controller of the Navy, Reginald Henderson, Brown devised a plan between them and Sir Charles Craven to take over an entire site belonging to Vickers exclusively for War Office tank work while allowing the Admiralty free rein over the rest. To assist, Craven put a substantial portion of the design department at Vickers at Brown and Henderson's disposal, and together they developed what Brown later described as a "most valuable asset." This was a glimpse of what industrial cooperation could achieve, two service departments and one industrialist reaching quick

[9] TNA, CAB4/27, 43rd meeting of the DPR, 30 September 1937.

decisions on re-organising and dividing a facility for the benefit of all sides, free from political and financial controls. That it took until late 1937 to develop, and even then, only because Brown could call upon his naval connections to assist the War Office, aptly illustrates the limitations of the subcommittee system that reported upwards to Cabinet.

Brown went on to call upon Craven, Weir, Lithgow, Balfour and Harry McGowan (from Imperial Chemical Industries) frequently from 1937. They helped with a wide range of problems – at one point in 1938 even suggesting clothing manufacturers for Territorial Army uniforms.[10] Brown remarked that:

> Amongst the many eminent industrialists who came in to help us at an early date, first and foremost stands out the name Lord Weir, whose sage advice had so long been available to the public service as industrial adviser … [He was] an outstanding example in the way in which such work could be done and for which his commercial contacts made him uniquely competent.[11]

Weir was later noted by Brown for arranging a "most effective unit" for the production of field artillery by a group of non-armament firms from 1938, of which his was the leader.[12] Among other things, he was also brought in by Brown to advise the Army Council, perhaps repaying the earlier faith Weir had shown by supporting Brown's move to the War Office.[13] In marked contrast to the opinion of his assistant, Turner, Brown thought that the "knowledge and ready assistance" of the industrialists "proved invaluable" in the rearmament effort.

Weir, like Lithgow with Fairfield and Beardmore, profited financially from his connection with Brown and the supply organisers. As a direct result of these new orders, income from the land division of the firm increased by £135,000 in 1939 compared with four years earlier. The scheme to create a small group of 400 parent firms which would educate others – first proposed by Weir in 1933 – put his firm in a strong position once rearmament began. Admiralty orders jumped from £445,000 in 1935 to £944,000 in 1937 and then to £1.9m in 1939 – between forty and seventy percent of total orders by value. Similarly, the firm's profits grew from £172,000 to £462,000 in the same period. Total sales quadrupled to £2.8m. His company undoubtedly did very well out of rearmament orders, but it did so because it was one of just a few already capable of making large quantities of the specialised equipment the service departments required.[14]

---

10 Brown, "Reminiscences," II, 197.

11 *Ibid.*, II, 214–215.

12 *Ibid.*, II, 183.

13 Reader, *The Weir Group,* 129.

14 All taken from figures in *ibid.*, particularly 134–135.

Brown also remarked that he owed an "eternal debt of gratitude to Craven," who "gave every possible assistance, both as regards design and production" to him and his department in times of need. Lithgow gave "ready and substantial assistance with gun production" and "placed all of [Beardmore's] substantial assistance at the War Office's disposal" (other than that already allocated to the Admiralty) and undertook considerable extensions to the Parkhead works for naval and field artillery gun manufacture. He also assisted in 1938 with what Brown described as a "very considerable expansion of production of bulletproof plate and gun forging capacity" at Beardmore.[15] Sir Andrew Duncan, ever more closely connected with Weir and Lithgow, offered "invaluable guidance on all questions of steel and raw material supply."

The common theme – perhaps with Duncan excepted – was the connection with SC III. These men at one point or another had distinguished themselves to Brown, and he was now making full use of his connections and their talents. More strikingly, in any defence supply matter or business acquisition before rearmament, one from this very small group was never far from it. The "web" of industrial influence always seemed to have Weir, Lithgow or Duncan near its centre.

Along with Henderson and Air Chief Marshal Sir Wilfred Freeman (Dowding's successor as the RAF member on the Supply Board), Brown developed a scheme to reach more agreements of the kind he had worked out with Vickers. To this end, he lunched with Craven and Lithgow regularly and formed an unofficial priority and liaison committee to settle the numerous competing claims on resources among the three services. Known somewhat obscurely as the "Boilermakers," this small group, meeting in a hotel restaurant, discussed and helped overcome the pressing inter-service problems and functioned, in the eyes of Brown, with "unvarying success" until the war. Amusingly, after the war began the Minister of Labour (a position created in 1940) Ernest Bevin suggested that the heads of production for all three services should meet regularly to overcome supply problems. Brown graciously accepted Bevin's wisdom, since in his own words "this already happened with the Boilermakers … [so] all that this meant was that Her Majesty's Government ended up paying the bill for lunch!"[16]

From this, Craven, Lithgow, Rebbeck (of H&W) and Henderson began to informally allocate shipbuilding capacity as a "shadow committee," and it would be to them Brown would go if he wanted use of some shipbuilding capacity that had been allocated to the Admiralty.[17] Of course, having three WSBC members working with the Controller of the Admiralty gave the WSBC even more information on likely requirements than they hitherto had, and it is likely through

15 Brown, "Reminiscences," II, 143.

16 *Ibid.*, II, 199–200.

17 *Ibid.*, II, 199.

this route that the WSBC was able to accurately anticipate future naval orders and to invite firms with the most spare capacity to place bids for them. Still, it seems that information flowed freely through clandestine channels from the CID to the WSBC, but rarely, if ever, in the other direction.

Like Brown's agreement for the plant at Vickers, these were more examples of leading industrialists offering their services willingly to the defence departments and reaching amicable conclusions in a short time frame that benefitted all parties. Tendering expenses aside, this was certainly a more efficient way of working through problems without getting bogged down in inter-departmental disputes. However, the Boilermakers need never have existed. They were only duplicating the work of the Supply Board, insofar as they aimed to allocate resources for air, sea and land around the available armament-making facilities. That Brown felt it necessary to bypass the CID committee structure to get things done without escalating the matter upwards to Inskip underlines the limitations of the system he worked with for so long.

Another aspect was how Brown and the industrialists viewed Inskip. Brown appeared to respect him but "always believed his appointment was political" and that he primarily existed to "curb the demands of service ministries and to bring them into line with the government desire for a more limited war."[18] In other words, he believed that Inskip was chosen to be the Cabinet's man in the CID, not the CID's man in Cabinet. Lithgow's relationship with Inskip has already been covered.[19] Weir is somewhat harder to read, but it is known that his relationship with Inskip noticeably cooled in early 1937 after Inskip decided to re-open the question of the Fleet Air Arm – essentially over whether the Admiralty or Air Ministry would control the development of aeroplanes for naval vessels – which Weir had previously advised on and considered closed. Weir bluntly told Baldwin he was not prepared to re-examine the matter, so when Weir found out that Inskip had written to Baldwin without consulting him, he was furious and threatened to resign from the government.[20] Although Inskip later apologised profusely, resignation was not a step Weir, with lengthy public service and loyalty to the government took lightly, so it is hard to believe the relationship was fully mended thereafter.[21] In sum, Inskip appeared to inhibit rather than facilitate the work of supply committees. Weir, Lithgow, Brown and other experts thus decided they were better off solving problems informally outside of Whitehall rather than dealing with the hierarchy of committees which led upwards to him.

---

18 BLH, TB, box 1, "Inskip Notes," undated.

19 GUAS, DC35/31, Inskip to Lithgow, 9 November 1936.

20 S. Greenwood, "Sir Thomas Inskip as Minister for the Co-ordination of Defence, 1936–1939," in P. Smith (ed.), *Government and Armed Forces in Britain, 1856–1990* (Cambridge, 1996), 165–166.

21 CCC, WEIR19/14, Inskip to Weir, 12 January 1937.

The Inskip Report and Financial Control, 1937–1939

The Treasury had stuck to Weir's principle that the "word of the man in supply" should carry and took a light-touch approach to the negotiation and scrutinising of defence contracts in 1936 and 1937.[22] This freedom allowed for men like Brown to get on with the business of rearmament. The most efficient way of doing so was to cut through the red tape of the CID's subcommittees, which were consistently hampered by constant referrals to other bodies for approval. The result was that Brown and industry operated informally by gentlemen's agreement in 1937.

The cost of rearmament projects was also escalating rapidly. In February 1937, the National Government had brought forward the Defence Loans Bill, which sought to raise another £400m for rearmament. This was only a taste of things to come. The Chancellor, Neville Chamberlain, succeeded Stanley Baldwin as Prime Minister at the end of May. Prior to this, one of his last acts as Chancellor had been to insist upon a "rationing" of future defence expenditure and to push for Inskip to review defence policy considering the burgeoning cost. Part of the rationale behind this move was Chamberlain's belief that Britain's ability to wage a long war would be reduced if the rearmament programme undermined economic prosperity. The theory was that the more prepared Britain was militarily, the more resources were diverted from other areas of the economy and the weaker she was financially – hence the demand that rearmament not interfere with the normal business activity of the country.[23] While on one level this was certainly rational, adherence to "business as usual" had the consequence of lessening the focus on industrial resources and inefficiently allocating them to civilian tasks.

On another level, this was also an attempt to re-establish a form of Treasury control over expenditure that had been slipping away. As Cabinet was aware of the need to be seen taking firm action on the excessive profits of armaments contracts, this was a review that it was only too happy to support. Inskip thus worked on an interim report on defence expenditure throughout the second half of 1937 and presented his first set of recommendations that December. In the short term, the Treasury's request had some effect: no new increases in the defence programmes were to be undertaken until Inskip had presented its findings.

Inskip ultimately concluded that total expenditure should "not exceed" £1500m in the five financial years that followed – although by February 1938 he had already revised this figure upwards by another £70m – and soon after believed that £2000m was a more realistic figure.[24] By way of comparison, the total for

22 Peden, *British Rearmament and the Treasury,* 46.

23 G.C. Peden, "A Matter of Timing: The Economic Background to British Foreign Policy, 1937–1939," *History,* 69/1 (1984), 17; and TNA, CAB23/87/10, CC 8(37), Meeting of the Cabinet, 17 February 1937.

24 Peden, *British Rearmament and the Treasury,* 41–42.

the five years to 1937 had been around £600m – or rather less than one-third of what Inskip believed would now be needed. The Treasury had previously stated its belief that around £1100m could be raised in that period through prevailing rates of taxation,[25] so there is perhaps an element of Inskip attempting to fit the square peg of rearmament into the round hole of the £1500m that could be raised through a combination of taxation and the £400m defence loan.

In terms of "reasserting" Treasury control over expenditure, Inskip's report had only a limited effect. The *Anschluss* in March 1938 and the diplomatic crisis that followed overtook any serious attempt to cap costs. The increasing pace of German aerial rearmament prompted Cabinet to agree in April 1938 that the Air Ministry could order as many aircraft as it felt industry could provide before 1940 – and once war broke out any previous ideas of limits became irrelevant. The Admiralty was on a marginally tighter leash in 1938, but still the First Lord of the Admiralty got some £10m *more* than he expected when agreeing the naval share of the ration that summer.[26] Thus, despite bringing forward the idea that there should be a limit, even if loosely defined, on what Britain could afford to spend on rearmament, the "rationing" of expenditure was not particularly prohibitive in the short term. This did not necessarily mean that Chamberlain's mindset had changed: in October 1938, he reasserted that finance remained Britain's "greatest asset" in a long war.[27] The problem was that a stage had been reached where Britain had little choice in the matter.

Moreover, in practice finance was less of a limiting factor than what industry could physically provide in 1938; the Treasury and CID both understood that even if the total sum over five years was adhered to the individual departments were free to front-load expenditure rather than spread it evenly, but that the level of spare capacity would mean that some of the money simply could not be spent fast enough.[28] However, in 1938 at least, there is some evidence that the mere existence of rationing as a concept prompted service departments to think about ranking expenditure priorities. Nevertheless, this still meant that the word of the "man in supply" was carrying, and it was in any case clearly not a return to the early DRC period in 1934 with Chamberlain attempting to revise the conclusions of the CoS and Hankey.

Indeed, it was at this point that another central tenet of government rearmament policy was abandoned. The CoS had been increasingly unhappy with the principle that normal business practice should not be interfered with. The heightening continental crisis convinced Cabinet to finally change tack in March 1938 and instruct (but not compel) the industry to give full priority to

---

25 *Ibid.*, 78 and 205.

26 *Ibid.*, 43.

27 TNA, CAB23/96/2, CC 49(38), "Cabinet Conclusions."

28 Gordon, *British Sea Power*, 258–260; and Parker, "British Rearmament, 1936–1939: The Treasury, Trade Unions and Skilled Labour," *English Historical Review*, 96/2 (1981), 311.

rearmament.[29] This was a significant – although strikingly late – turning point, for it finally gave authorisation to all three services to place contracts (in particular the "educational orders" that the Advisory Panel advocated) in industries that were normally engaged on civil rather than military goods. In other words, the brakes were only fully turned off three years after the PSOC and industrialists believed necessary. Admittedly, the effect was less significant for the Admiralty, for it already drew upon a group of established manufacturers and had technical bottlenecks that could not easily be solved by education alone. It did, however, impact upon the work of Weir and Lithgow, who were by the summer of 1938 busy advising on new sources of supply from civilian manufacturers – including, as already noted, everything from uniforms to field artillery.

From the beginning of 1939, what little Treasury financial control remained broke down completely. Estimates for that year jumped by almost £200m in a matter of four months, with the Cabinet swiftly agreeing to a series of additions to existing defence programmes without consulting the Treasury.[30] Throwing ever-increasing sums at the problem was not a particularly good solution in 1939, however, for there was precious little spare capacity in established industries to cope with demand. As such, the period between February and the declaration of war was as much about building whatever could be built with the capacity available as it was about meeting strategic priorities. The Admiralty wanted the *Fiji*-class cruisers to have seven double-gun turrets, but it had to make do with four triple-gun turrets instead, for there were insufficient mountings to complete even the guns on the *Dido*-class, its predecessor.[31] Naval shipyards warned of a lack of armour plate to complete vessels, prompting the Admiralty to issue instructions on how to store the incomplete ships until the parts and components arrived.[32] Moreover, in 1939 it ordered seventeen fishing trawlers in February, and would eventually order 200 more. At just 500 tons, these trawlers were a fraction of the size of even the smaller destroyer classes and were armed only with one three-inch gun, in many cases reclaimed from scrapped vessels – essentially technology from the 1890s, firing a shell one two-hundredth the size of a modern battleship – and highly rudimentary armour.[33]

Although of very limited use, these naval trawlers were to be used on minesweeping and harbour defence duties to allow larger ships to be deployed elsewhere, and, most importantly, could be built to a simple standardised design in non-specialist and non-naval yards. Naval trawlers had first built to an Admiralty emergency programme in 1915 on an *ad hoc* basis because of the

[29] Still, this stopped short of giving government controls over the means of production. See Gordon, *British Sea Power,* 264.

[30] Peden, *British Rearmament and the Treasury,* 44.

[31] TNA, ADM1/9401, "Eight Thousand Ton Cruisers – Designs," 1936.

[32] GUAS, UCS1/1/2, John Brown Papers, "Minutes of Meeting," 17 February 1938.

[33] P. Elliot, *Allied Escort Ships of World War II: A Complete Survey* (London, 1977), 293–295.

demands of the ongoing Great War. However, the fact that so many were once again ordered at such a late stage and not seriously considered during earlier defence schemes before 1938 suggests desperation on the Admiralty's part to get whatever it could from whomever it could. Moreover, it shows how well placed the earlier concerns of Lord Weir were, and, most aptly, what Harold Brown later described as "a trickle ... then a flood" of contracts. In other words, not enough naval work to increase or even sustain capacity before 1936, and then, with the political situation deteriorating, too much to cope with by 1939.[34] This was the manifestation of the failure of policy: these problems had been spotted years earlier, but had not been solved by 1940.

Thus, when the Second World War began, most of Britain's major rearmament-era naval orders were not yet complete. The first two *King George V*-class ships, laid down on New Year's Day 1937, were not commissioned until October 1940 and January 1941, respectively. The final ship did not enter service until June 1942.[35] Similarly, none of the post-Treaty aircraft carriers, and very few of the cruisers, were commissioned before the summer of 1940.[36] Indeed, the majority of combat-ready ships – particularly the larger vessels – when war was declared were comparatively old; *Nelson* and *Rodney* were the youngest of all the battleships or battlecruisers, yet they had been laid down back in 1922. The rest were relics of the Great War. The youngest of Britain's complement of heavy cruisers was laid down in 1928, while no ships of the first post-Treaty light cruisers, the *Dido*-class, were yet finished. Only one aircraft carrier out of the Royal Navy's seven, the *Ark Royal,* had been ordered since 1918. In terms of carrier development, the age of the vessels correlated well with their collective fate: no pre-1936 aircraft carrier survived the war (five were sunk and two were deemed obsolete for combat and retired before 1945), while every post-1936 carrier did.[37] Rearmament planning from 1932 did not mean a new, modern navy by the conclusion of 1939.

Structurally, the system was also changing. In 1939, the CID was wound up and replaced by a new Ministry of Supply to coordinate the work of all three defence arms. This did not come entirely out of the blue. The Royal Commission suggested the idea (Hankey buried it), and Churchill and others had made significant speeches in the previous two years arguing for a new Ministry of Munitions,[38] like the organisation he had headed during the Great War. Churchill claimed the CID had failed to have the "wit and foresight [to] ... make the

34 Brown, "Reminiscences," II, 153.

35 Buxton and Johnston, *Battleship Builders,* 42–44.

36 For an overview of the larger ships under construction and in commission as of 1942, see McMurtie (ed.), *Jane's Fighting Ships,* 16–45.

37 This is not to suggest that the lack of armoured decks contributed to all the sinkings – *Ark Royal,* for instance, was destroyed by a torpedo.

38 *The Times*, 11 July 1937 and 1 June 1938 are two such examples, although more exist in the House of Commons.

preparations in the factories which would … have in 1933 and 1934 enabled the whole business of making weapons get into its stride two years earlier."[39] Indeed, the quote from the *Gathering Storm* which was used to introduce this book can be read as another, later, example of Churchill's criticisms of supply organisation. However, for all this, Churchill was oddly quiet in the matter between 1933 and 1936, so he seems to have retrospectively changed his mind.[40]

Moreover, while Churchill's point that the pace of change could have been quicker is partially valid, he failed to account for the political difficulties and constraints of the day, or that Weir, Sir Arthur Robinson and Inskip had considered the question in 1937 and concluded it would have been politically very difficult, if not impossible, to create a new ministry that had the power to compel industry to hand over production to the state.[41] Furthermore, as the position the Admiralty and War Office took in the Supply Committees in 1933 and 1934 illustrates, the defence departments all had different opinions on how to organise production and were averse to handing over responsibility to a central body.[42]

This impasse did not break down until the spring of 1939 when the German *Wehrmacht* occupied Prague. The need to put the army on an immediate war footing accelerated the issue of state control of industry, and the bill came into force creating the Ministry of Supply in August.[43] The new organisation was not as radical as one might at first believe, however. Despite more powers over industry, it kept the principles of inter-service cooperation that had been built up since the 1920s and left the Supply Board (including the Supply Committees) and Admiralty procurement channels untouched.[44] Though presented as a wholly new and fresh organisation ready to tackle the demands of war and a departure from the "talking shop" that it replaced, the new Ministry was not the organisation that Hankey, Weir and the Admiralty rejected. It was essentially a tweaked PSOC with enhanced powers and was something that could and perhaps should have come into existence years earlier.

In sum, the system of supply organisation before 1937 did not fail to spot, but did fail to remedy the industrial bottlenecks that existed in the naval armaments industry. It took until 1937 to properly begin the rearmament process and until the middle of 1938 to fully direct the industrial effort towards defence. Britain was better prepared than it might have been, but only thanks to the work of men like Lithgow, Weir, and Craven in industry and Brown and Henderson in the CID, essentially acting without official consent to better

39 *The Times*, 1 June 1938.

40 Gordon, *British Sea Power,* 250.

41 TNA, CAB2/7, Discussion of "Report of Committee on Supply Organisation in War," 309th meeting of the CID, 3 December 1936.

42 TNA, CAB60/42, PSO(SB)457, "Functions of the Supply Board in War and Means by which They may be Exercised," 4 October 1934.

43 Shay, *British Rearmament in the Thirties,* 273–274.

44 Cmd. 6034, *Ministry of Supply Bill*, May 1939.

organise resources. Nevertheless, despite thinking, investigating and reporting constantly since Manchuria and offering numerous sound recommendations and innovative suggestions, the lower-level planning machinery was poorly utilised by the top-level decision makers. The issue of state control of industry thus misses the point: it should not have been about compelling industry to turn over production to the government in 1938, but government acting years earlier to utilise the concepts of shadow industries, educational orders and financing re-tooling so that this crisis point was never reached in the first place. While the distinct political economy which prioritised planning in the 1930s was not without benefit, its adverse side effects for industrial preparation are impossible to ignore.

Supply Organisers at War: The WSBC and the Admiralty, 1939–1941

The spate of new contracts awarded in 1937 and 1938 played into the hands of the WSBC, which recorded handsome profit margins on naval orders. What follows considers the fortunes of the WSBC during the war, though a more detailed account of private industry during the first two years of war lies outside of the scope of this book. The reason the narrative concludes in 1941, not 1939, is that it offers a natural conclusion to the history of the WSBC, as well as a point at which to take stock of the development of supply organisation and industrial involvement within the British state across the preceding two decades. Moreover, as far as the private manufacture of naval arms was concerned, 1939 was not an end to rearmament; it should be recalled that Inskip's attempts to put a total cost upon equipping British defences was based on a calculation of the financial years 1937–1941.[45] Similarly, Weir and the DPR believed that 1941 was the point by which industry could complete the defence scheme which had been first outlined in 1936.[46]

For private industry between 1939 and 1941, it was the business of rearmament that dominated proceedings – in the main working on finishing or fitting-out ships which had been ordered some years earlier. Profits remained healthy, despite a new profits tax introduced in 1939. This attempted to claw back anything above "average" profits in recent years of peace – which, for the WSBC, had been far higher than normal anyway.[47] But the main development had little to do with this measure. Instead, it was the Admiralty that from the end of 1940 began to take a much keener interest in the prices it was paying for equipment. This, for the first time, threw light on the illicit dealings of its industrial partners.

---

45 Peden, *British Rearmament and the Treasury,* 42.

46 This scheme had been substantially expanded in 1938 and 1939.

47 W.A. Thomas, *The Finance of British Industry, 1918–1976* (New York, 1979), 231.

As Treasury control broke down, Admiralty control over prices and profits was stepped up from 1939. There has been an oft-held assumption – by contemporaries and historians alike – that the Admiralty had the most robust contract structure and was better prepared to deal with profiteering, for it utilised a group of "experienced firms, with a long record of Admiralty contracts behind them" and possessed its own Royal Dockyards which "set a standard for costing purposes."[48] In fact, the opposite was the case. The Admiralty possessed the basics of an effective structure, certainly in comparison to the *ad hoc* nature by which the War Office negotiated contracts, but naval firms used their ample experience and track record to their advantage while avoiding detection. The existence of the Royal Dockyards gave only the illusion of strength, as both firms and the Admiralty knew that effective comparisons between privately built and Dockyard-built ships were almost impossible to achieve with any accuracy.

It should also not be assumed that the Ministry of Supply was more robust than the CID/PSOC organisation it replaced. In truth, the Admiralty spent most of the period between 1922 and 1939 in a state of complacency: it possessed the ability to rigorously inspect and compare the books of firms working on a government contract, but almost never did so. Despite the National Government's outward commitment to the prevention of profiteering, the systems in place to limit its effects were woefully inefficient in 1939. The National Defence Contribution – a five-percent tax on all profits – barely dented the profits recorded on Admiralty vessels, while the later Armaments Profits Duty – a sixty-percent tax on the profits from rearmament contracts – was only in use for four months in 1939 before being replaced once war broke out.[49] As such, for the entire period of the WSBC's existence from 1926 until the war, they were not effectively audited by the Admiralty or Treasury.

From surviving records, the earliest indication that the Admiralty intended to investigate the practices of the WSBC was in November 1940. In a letter to John Brown's Clydebank yard, the Director of Navy Contracts, E.C. Jubb (the same man who had in 1936 been so confident that competitive tendering and the existence of the Royal Dockyards provided adequate protection against profiteering)[50] queried the overhead charges given by John Brown for a *Roberts*-class monitor, which had been tendered for that May. John Brown claimed that the "disturbance of work owing to the necessity of transferring workmen to emergency work" [i.e., war contracts] necessitated an extra charge being included in the tender. The Admiralty believed that owing to John Brown's "current large output" and high levels of staff, the forty-percent

[48] Peden, *British Rearmament and the Treasury,* 50.

[49] Rollings, "Whitehall," 530–531.

[50] CUL, MS Templewood, part 9, file 5, "Admiralty Contracts, Prevention of Profiteering," 17 July 1936.

increase in overhead charges should have been much lower. Jubb thus urged the company to "reconsider the amount quoted … with a view to a very substantial reduction being made."[51]

By the following February, the Admiralty had written to all WSBC firms noting that in light of the "very considerable expenditure of public money now being incurred on the building of warships" it was requesting that firms allow an accountant to visit to undertake investigations into costs. While it did not propose alterations to contracts, this was unquestionably the start of a much more rigorous attempt to gain a handle on how private manufacture had arrived at figures quoted for Admiralty contracts.

Typically, the WSBC's communication networks meant its members kept abreast of developments. In April 1941, for instance, J.A. Milne, a director of J. Samuel White, wrote to Crease at the WSBC to circulate to the other members the details of the Admiralty accountant's visit to White's Cowes yard. In it, he noted the contracts which were being investigated, the duration of his stay, and the files he requested. Milne noted that the accountant "confined himself to asking for half a dozen invoices … [and] finding a satisfactory system of internal check in operation, made no detailed investigation of the figures in the Cost Ledger." Tellingly, he then reported that, as the "cost and financial accounts reconciled," it was "probable that the fact that this was possible had some influence in his decision to make no detailed investigation of the actual costs" and thus "he made no enquiry into the selling price or make up of the tender." Finally, Milne expressed his concern over the Admiralty discovering the WSBC's "special fund" which the members had been adding to peacetime orders to assist builders or to help win foreign warship work. This totalled £1500 per ship on the contracts the accountant had seen. Milne believed that if the "special fund were disclosed … and the Admiralty asked for explanations, the Shipbuilding Conference should furnish a general statement on the subject, rather than leave the individual Warship Group firms to evolve the explanations."[52] In other words, Milne knew his firm had a lot to hide, and how fortunate he was that the Admiralty accountant did not uncover evidence of price-fixing.

Still, this soon evoked panicked responses. Crease first understood that the investigations would not involve ships started during peacetime, but now believed these would be treated as war profits and feared that the results would "have a very disturbing effect throughout the Admiralty." Moreover, he believed that "there [was] very little chance that knowledge of the results would be confined to the Admiralty" and as such concluded that they would "come before the Select Committee on National Expenditure and thus to the Treasury and Parliament."[53]

51 GUAS, UCS1/21/17, Jubb to Piggot, 16 November 1940.

52 Quotations are from GUAS, UCS1/21/17, Milne to Crease, 5 April 1941.

53 GUAS, UCS1/21/18, Crease to "All Warship Group Firms," 18 August 1941.

His chief concern, therefore, was "whether any action can be taken by the firms to improve the situation as it now exists" and "for firms, where necessary, to amend their tenders in the light of experience that has now been gained of their costs under war conditions."[54] In August, Crease raised the idea of offering some form of rebate to the Admiralty – probably the clearest sign yet that the group accepted guilt – although Stephen Piggot of John Brown believed that such a scheme "may be construed as being done solely due to the results ascertained from the cost investigation."[55]

Another suggestion, this time from Cammell Laird, was to open to the Admiralty the books of all ships going back to Washington, with the hope being that the lower profits – and indeed losses – in the 1920s and early 1930s would offset the later gains made during rearmament.[56] Cammell Laird noted that the average profit on all ships since *Rodney* in 1922 was a much more modest 12.5%, particularly because of losses incurred on earlier orders.[57] Crease suggested that this should be put before the committee first, and that Cammell Laird should not make this offer to the Admiralty before there was a chance for "general consideration" because "such action on your part might practically necessitate similar action by all the other firms."[58] Moreover, he believed such an invitation for the period since 1923 would:

> Include amongst other things all the arrangements for pools [rotas], which persisted in the case of smaller ships, all the arrangements for the *York*-class cruisers in 1937, and the business with the purchase of the Dalmuir rights in 1934 and 1935. The possibility that any of these matters might be disclosed as the result of carrying back the investigations would give me great uneasiness, and the result might well be to increase our present difficulties.[59]

Thus, while it was recommended that while a scheme to get the Admiralty to investigate ships that yielded lower profits was desirable, this was perhaps not the way to go about it.

In October, the Admiralty – while not threatening to go to Parliament (presumably through the embarrassment it would also cause its own contracts department) – did ask for more detail of how costs were arrived at, and provided notifications to firms for the ships they proposed to investigate. From this, the WSBC attempted to push the Admiralty towards ships with lower profit yields.

54 *Ibid.*
55 GUAS, UCS1/21/18, Piggot Notes, 20 August 1941.
56 GUAS, UCS1/21/18, Johnston to Crease, 11 September 1941.
57 GUAS, UCS1/21/18, Crease to Committee, 17 September 1941.
58 GUAS, UCS1/21/18, Johnston to Crease, 11 September 1941.
59 *Ibid.*

In the case of Harland & Wolff, the aircraft carrier *Formidable* was selected, and Crease asked other firms:

> … whether any benefit would result from investigation of *Illustrious, Victorious* or *Indomitable* [the other three ships of the same class in other yards] which are also available. It must be observed that the Admiralty stated that we must not select or nominate individual ships. We could therefore only ask that another carrier be investigated.[60]

This was clearly linked to Crease's suggestions in August and September and highlights the WSBC's fear of the Admiralty's reaction if it uncovered an abnormally high result. The same was true for the *Hunt*-class destroyers. Crease believed that "we might ask for the 3rd and 4th groups to be investigated," for these "would show considerably reduced profits." He noted that with profits of 35.4% and 33.7%, respectively, "none of [the first and second] group figures are very helpful to our case." In an update the following week, he reported that he "gathered the results on *Formidable* will not cause embarrassment, but [was] doubtful about the other items."[61]

In sum, the evidence for WSBC price fixing is plentiful and undeniable. That it took a major war with substantially tightened rules to uncover systematic overcharging – and even then, only the excessive profits recorded immediately before and during the war itself, is testament to the sophisticated way in which WSBC firms devised their tendering scheme. By the end of the following year, the Admiralty had a much firmer grip on prices. In 1942, Swan Hunter won a contract for a floating dock valued at £150,000, and had the cost investigated by Jubb. He then made them an offer of six-percent profit – a vastly reduced sum – to which Swan Hunter protested and asked for ten percent. The final settlement was seven percent. Writing to John Brown, Swan Hunter's chairman noted despondently that "it was rather a struggle, and I am afraid you may not be able to get any settlement at a higher figure, especially in light of their present attitude."[62] There never was any parliamentary enquiry to force a rebate for the profits generated in the war, nor was there a full enquiry into tendering going back throughout the years of disarmament.[63] While this was probably some consolation to private industry, it was nevertheless clear that the era of the profiteering cartel was over.

It remains now to return to the fortunes of the small coterie of industrialists. Surprisingly, detailed and substantive accounts of their activities during the war are relatively sparse, although Weir, Lithgow, Duncan and others all

60 GUAS, UCS1/21/18, Crease to "All Warship Group Firms," 13 October 1941.

61 All quotes are from GUAS, UCS1/21/18, Crease to "All Warship Group Firms," 9 October 1941.

62 GUAS, UCS1/21/18, Swan to Piggot, 23 September 1942.

63 C.W. Miller, "British naval armaments, cartels, and defence planning between the world wars," *Entreprises et histoire*, 85/1 (2016), 86.

moved quickly into senior supply positions in the wartime government. Weir was Director-General of Explosives, Lithgow took on the role of Director of Merchant Shipbuilding and Repairs at the Admiralty, and Duncan held the position of Minister of Supply for all but one of the six years of conflict.[64] The biographers of Weir and Lithgow (Duncan has no biographer) rather gloss over this period, stressing only the excellent service they gave to their country. In the case of Weir, his personal role is almost entirely omitted, and just thirteen pages are given to the fortunes of the entire company between 1939 and 1945.

However, Brown's secretary at the War Office, Sir George Turner, who held rather dimmer opinions of Weir than most in the last years of peace, also kept a diary of supply organisation during the war itself that offers a rare – and critical – insight into the role of these "experts" within government. On Duncan, the man appointed as Minister of Supply at Weir's suggestion, and someone with "a big name in business, he said

> … how he got his name becomes a mystery when one contemplates his performance as a minister in charge of a department for which he was supposed to be specially fitted. He is afraid of the Prime Minister but at the same time a bully to his subordinates …[65]

Duncan initially did not last long in his position and was asked to resign (he returned a year later, however). Turner was clearly delighted by this outcome, as he believed that "Duncan wrecked the ministry by his vacillations and constant changes." When he was replaced (with Lord Beaverbrook), Turner noted that Weir "went off to Scotland, and declared his intention not to return," followed by Lithgow, which he put down to Duncan's previous career on the Clyde, with Weir and Lithgow as his original masters.

Turner summed up by remarking that Weir had a track record of "jobbing" his friends into jobs in government, including his brother (James), who was brought in to assist with tank production. Here, Turner rather cuttingly remarked that his only qualification for that job was "his relationship with Lord Weir." Perhaps most damningly of all, Turner concluded with the assertion that

> however able these "imported ministers" may be, few of them really seem to have the necessary background … The fact is outside business experience rarely covers so wide a field. It puts too much emphasis on cash values. This is one of the weaknesses of appointments from outside the parliamentary team.[66]

64 Reader, *The Weir Group,* 134; and Reid, *James Lithgow,* 193 and 203.

65 BLH, TB, box 2, Turner Diary, 2 July 1941.

66 *Ibid.*

In short, Sir George Turner may have been a lonely critical voice – perhaps the only such figure on record to dissent in the entire period since 1933 – but this does not necessarily mean he was entirely wrong. His comments in 1941 thus offer a vantage point from which to evaluate the central theme of this book: the role of industry, and industrial influence within the state, in disarmament and rearmament.

Conclusion and Retrospective

While events after 1939 lie, in the main, outside the remit of this book, the general questions throughout have been to consider the role of industrialists and the private naval arms industry within supply planning, and to assess how they influenced, and were affected by, government policy. More generally, it has sought to examine decision-making processes and the effectiveness of the supply planning machinery in the CID and related bodies throughout their development in the 1920s and 1930s. Therefore, the early phase of the war and the culmination of years of planning and organising (and price-fixing) serve as the best place to conclude this work, rather than the outbreak of war itself.

Bearing this in mind: there are some overarching points to be made. The first concerns the role of an industrial elite within the state. This book has shown that in the case of naval arms there was both an elite group of "insiders" and a defensive ring of "outsiders" before 1936. There was no "web" of well-connected individuals conspiring with government to profit from the standard kind of "Military-Industrial Complex" that subsequent theories have sought to find. In the 1920s, the Admiralty and private industry certainly shared interests in the preservation of capacity, and as such jointly devised methods to protect facilities as best they could in the circumstances. This was, however, the result of a lack of power and influence, not evidence of it. At any rate, the schemes were small-scale and extended no further than a work-sharing rota or minor subsidies. Indeed, before Manchuria in 1931, the Admiralty struggled to articulate the need for more or newer vessels in the face of Treasury opposition, leading to the decline of the Royal Navy as an effective global fighting force and to the unquestionable decline of private manufacturing capability.

In the case of the "insiders," their story was in some ways not unlike the 1920s Admiralty-industry relationship that brought about the WSBC. Again, it was not a vast shadowy network of industrialists influencing state decision making. The Supply Board turned to a handful of tried and trusted individuals, names which appeared and re-appeared in a variety of situations between 1920 and 1940. The Advisory Panel of Weir – by far the most influential – Lithgow, Balfour and some close associates, particularly Andrew Duncan, gained access to the corridors of power in a way totally unlike the overwhelming majority of their peers. However, unlike the work-sharing rota, these "insiders" were well connected, influential and highly respected. Much of what later emanated from the PSOC and, in Weir's case, the DPR, can be attributed to the advice these men gave willingly and freely, not least the "shadow" armaments scheme and the concentration of financial resources in a nucleus of established firms (which, of course, they all represented).

The key turning point was Manchuria. While it did not alter the balance of power between Admiralty and industry, it was nevertheless a critical point in defence planning between the wars – more so than the "beginning of rearmament" supposedly brought by the formation of the DRC. It was Manchuria which forced the new National Government to think seriously about the *possibility* of naval conflict in a way that had been unimaginable for most of the preceding decade. The Washington and London treaties – which had balanced the competing demands of the United States and Japan at the expense of scope for further British naval construction – left the Royal Navy underprepared for action. The National Government did not have the financial means or political will to take a decisive step, but it did allow for the revocation of the Ten Year Rule and the move towards a middle ground where Admiralty and Treasury disputes could be placed to one side while a full investigation into the condition of private industry was undertaken. It was essentially a stalling tactic to balance competing political, strategic and financial demands while fully satisfying no one.

The result of Manchuria was, ultimately, the rehabilitation of the PSOC. Despite being formed to succeed the Ministry of Munitions, it was a largely forgotten committee during the 1920s. In the period that followed, though, it and its subordinate bodies playing a far more significant role, with direct orders coming from the Cabinet to fully examine the "position of the private armaments industry in Britain's system of Imperial Defence."[1] This was essentially the first step in the collaboration with leading industrialists which lasted well over a decade into the Second World War.

This collaboration was about information-sharing between two groups with much to gain from each other, rather than about preferential deals, contracts or kickbacks. The industrialists involved certainly had much to offer the CID and provided thoughtful and innovative solutions to supply problems in the critical years between 1933 and 1936. Very few others could match this level of integration, and with the *possible* exception of Sir Harry McGowan's work in explosives at ICI, no other businessmen served for so long or from such an early point as Weir, Lithgow and Balfour. Certainly, none matched Weir's seniority and influence. Although Vickers has often been the principal focus of previous studies of the private arms industry, Craven, Vickers' managing director, had nowhere near the same level of access to the defence policy machinery before rearmament – and only joined as an advisor on shipbuilding in the summer of

[1] TNA, CAB60/13, PSO 357, "The Position of the Private Armaments Industry in British Defence," 8 March 1933.

1937. In short, Weir's service to the supply planning machinery of the British state in the 1930s was unrivalled.

This did not mean that Weir and his associates suddenly stopped being businessmen; when opportunities for influence and financial gain arose they usually grabbed them with both hands. Despite repeatedly claiming personal disdain for their role and justifying their work on the grounds of national security, the Advisory Panel found involvement within the CID and the information that came with it a powerful draw. They may have threatened to resign from time to time, but ultimately all served more or less continuously as advisors and coordinators until 1945.

These men *did* indeed care passionately about British defence and the health of British industry, but membership of the CID's subcommittees also allowed them (more) money and power. Lithgow expanded his business empire substantially, taking controlling interests in two major armament concerns despite no previous interest in such a line of work. Moreover, he bought Fairfield when its own directors felt the business outlook was hopeless and when no one else appeared interested, and he purchased Beardmore from the Bank of England through his connections to Duncan and Norman while infuriating Lord Invernairn in the process. Weir, like Lithgow, benefitted from this arrangement, although perhaps more through influence – his full membership of the DPR being one such example – than wealth. He did, however, win several contracts for artillery and like Lithgow was paid handsomely to retool his factories for wartime usage once rearmament began.

This was a relationship that cut both ways. Industrialists undoubtedly helped to preserve facilities, find new sources of supply, and set up British industry for a more rapid transition to a war footing. In return, they were provided with information – not financial kickbacks or preferential contracts – which included likely bottlenecks, tendering procedures and key firms. They then acted like rational businessmen. It should be remembered that this happened in a period when the National Government did *not* want to order substantially more ships or weapons, and it thus was quite unlike the scenario dreamed up by conspiracy theorists of a government working with industry to provide each other with monetary benefits. While this had elements of a "complex" – both sides needed each other and worked closely together – it existed without any preferential financial transactions. It was a relatively optimal situation where, without any budget, the PSOC managed to point Lithgow towards the capacity both it and the supply subcommittees were concerned about losing. They may not have bargained with Lithgow's later behaviour and his threats – a favourite tactic – to close plant down if he did not get his own way, but the arrangement generally worked well for both sides.

"Outsiders" responded in an entirely different way and with different results. While Lithgow and Weir had influence of a sort, the WSBC members banded together to increase the price for vessels because they did not. This was a profiteering cartel by any measure, or in the argot of the period a "ring."

Although it ensured no more than a bare minimum of survival before 1935, thereafter it reaped significant profit margins that eventually prompted Admiralty intervention. R.P. Shay has called armaments profiteering "the final manifestation of the breakdown in cooperation" between firms and the state,[2] but such a view does not fit with the evidence presented in this book. Profiteering was a product of neglecting industry in the 1920s, not a failure to cooperate in the later 1930s. The tools to prevent profiteering always existed, but they were not used because of Admiralty complacency and later the fear of derailing rearmament.

Like the Advisory Panel, this was an arrangement (albeit with unintended results) that over the period benefitted both the firms involved and the British state. Industry preserved capacity for the state without a need for direct subsidy or increased construction schemes (which, if funding had been available, the Admiralty was willing to provide), but it milked the rewards for some years afterwards. There is little doubt that this was a group of firms pushed together into a defensive ring, and having failed to find alternative sources of finance (either through subsidies or new business ventures) understood that they could not influence or effectively lobby the governments of the 1920s on matters of naval policy. Like the Advisory Panel, this scheme yielded benefits for its members, but also at times broke official rules and behaved illegally. Nevertheless, the truth behind its operation does not fit conventional pejorative descriptions of military-industrial complexes.

Definitions of such complexes have tended to stress good relationships with the major manufacturers, lucrative contracts and deals between government and industry. In the case of the WSBC, very little of this existed. This decision-making process – taken here to include first the 1920s Admiralty-Treasury negotiations over spending and later the PSOC and its subordinate bodies – has been the other major element of this book. The argument presented here is that the advisory structure of supply organisation that Weir helped create in the 1920s was broadly adequate for peacetime but was hindered by a lack of teeth to make necessary changes and, most importantly, by the inertia of top-level defence and supply policy between 1934 and 1936. Moreover, had a sensible set of compromises been reached in the 1920s that allowed for the preservation of naval capacity, the issue of cartels and profiteering might never have occurred. Of course, one might well argue that the costs of doing so versus the extra costs of warships through profiteering between 1937 and 1940 may have cancelled each other out. Nevertheless, it is suggested here that the behaviour of industry and industrialists can be explained in part by the failure of top-level supply *policy*, not the failure of supply organisation within the CID, which spotted – but could not remedy – the problems before rearmament began.

Until 1934, one can at least understand the constraints on policy which existed and hampered the work of the PSOC and Supply Board, and should

[2] Shay, *British Rearmament*, 290.

therefore not be too hasty to criticise the CID or characterise it as a "talking shop" in the 1920s and early 1930s. Indeed, one of the great advantages of such a system was that it got different service departments to the negotiating table where they talked through competing visions and claims, and came to reasonably amicable conclusions. This system was sufficiently successful and desirable that it not only survived unaltered between 1924 and 1939 but was transferred wholesale into the new Ministry of Supply during wartime. Despite these successes, in terms of rearmament and war preparation, the time was ripe for action in 1935, but Baldwin and then Chamberlain baulked.

Rather than acting, they leaned on the advisory structure and formed more committees. While one of these – the DPR – elevated Weir to being the most influential industrialist in the country, it also caused work to be repeated and questions already answered to be posed again. In this light, Inskip's appointment in March 1936 was more of the same. Conciliatory and willing to take advice, he typified the National Government's approach. Ironically, the general failure to compromise between Admiralty and Treasury in the 1920s was replaced with too much compromise, over too long a period, in the post-Manchurian CID of the 1930s. Of course, the naval treaties were still in force until December 1936, but the fundamental point is that a path existed to ease the identified bottlenecks before they became prohibitive, and this path was not taken. Indeed, the year following Abyssinia witnessed a repetition of earlier investigations by a totally new committee, instead of a period of building upon the foundations set by the PSOC. In short, in 1935 and 1936 the National Government listened, but when the time came, it did not learn.

Perhaps the most damning conclusion here is that, for all the planning, Britain was not ready to rearm in 1936. Much has been made by historians of the spare capacity in naval shipbuilding and Britain's strong "warfare state."[3] But this book has argued that in some key respects, particularly with regards to the bottlenecks of guns, forgings, mountings, armour and skilled labour, Britain was remarkably underprepared and simply could not rearm fast enough: ships cost significantly more, took much longer to finish, and had to be supplemented by fitting-out smaller vessels bought wholesale from non-naval yards. This was at a time when the private naval armaments industry was supposedly geared towards maximum speed and efficiency. That this was not pre-empted in earlier years' points to a failure at the top-level of policymaking, and a failure to prepare industry in advance of rearmament. These were, after all, not unknown or unexpected issues. N.H. Gibbs has argued that Britain "made the transition from apparently ill-prepared peace to total war smoothly and rapidly" in the sense that this meant the "maximum mobilisation of human and material resources."[4] This book has taken a different view: Britain took some sensible steps and made

---

[3] Edgerton, *Warfare State,* 35–39; and Packard, "British State-Industry Relations," 32.

[4] Gibbs, *Grand Strategy,* 782.

valuable preparations during peace, but the industrial preparation for war was certainly not as smooth or as rapid as it could have been.

A lack of money was not the main issue, at least not after Weir recommended that "word of the man responsible for supply must carry," but there is little doubt that a very modest increase in expenditure between 1925 and 1935 would have eased the constraints that existed after 1936 – in industrial terms, this was Britain's "lost decade." There is a common misconception that the Royal Navy was starved of funds *during rearmament,* but this is not the case.[5] Simply put, Treasury control took far too long to break down. Blame here can be reasonably directed at Cabinet for failing to heed Admiralty warnings, or for ignoring recommendations over the desirability of a steady replacement programme. Even during the years of the DRC's first reports, the standard Treasury response was to review findings, add up the figures, think about the implications, and present a conclusion some months later. This allowed the sting to be taken out of demands for increased funding (some of which was, in fairness, unreasonable), but this arrangement persisted for too long. There were, of course, understandable objections to "interfering with industry," but the problem was not interference or compelling industry to turn over production to armaments. The National Government did not separate investment in private capacity, either through direct funding or educational orders, from the undesirable aspects of control over industry. Moreover, the expansion of facilities and an increased pace of replacement would *not* have contravened the Washington Treaty (which was signed and adhered to with the best of intentions), and thus throwing money at the problem from 1937 did not make up for earlier mistakes. Given the long lead time, these failures are difficult to excuse.

Indeed, one can perhaps suggest that in the 1920s the formation of the WSBC was crucial to ensuring that naval rearmament was not even further behind by 1939. In addition, supply planners identified problems and utilised industrial expertise to devise methods to overcome them. The overriding problem was that it took far too long for the ideas to make their way through the advisory and subcommittee structure and take hold in the top echelons of government. On one more level then, the "outsiders" in the WSBC were not so different from Weir and Lithgow as "insiders:" both groups played to their strengths for self-preservation and the betterment of their businesses, and both had costs and benefits to the British government.

One can therefore conclude by once more returning to Turner's diary. In 1941, when Weir, Lithgow and Duncan's wartime powers were reaching their collective peak, Turner ended this section of his diary by lamenting that "there are unfortunately at least a few cases in which men of business take government appointments for very good business reasons. Even the kindliest critic may be inclined to wonder how far the firms are getting something in return. Patriotism

---

5 Gordon, *British Sea Power,* 259–260.

is not enough to explain all of this."[6] This book has argued that while this was not necessarily "unfortunate" for British supply organisation, for Weir, Lithgow and Duncan, patriotism and profit were, indeed, inseperable.

6 Turner Diary, 16 January 1941.

Appendix 1

Chart of Builders

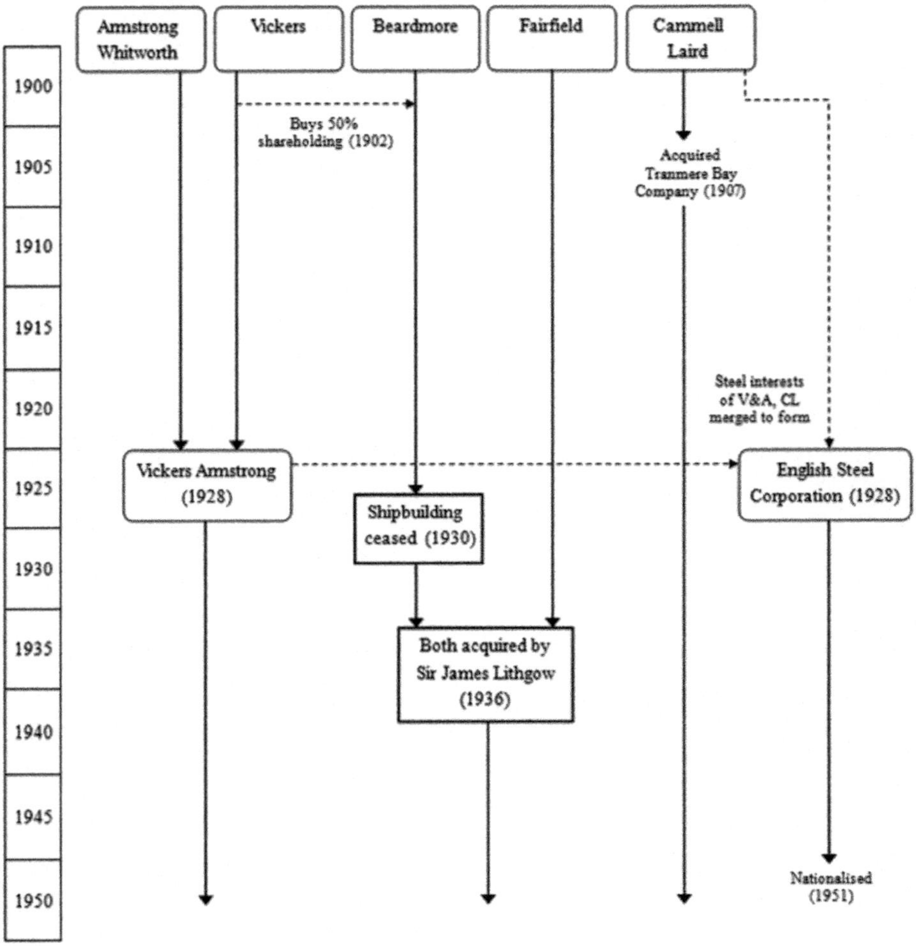

Partial histories of some major naval arms manufacturers, 1900–50. A dotted line denotes a partial sale or spin-off. Compiled from Johnston & Buxton *Battleship Builders*.

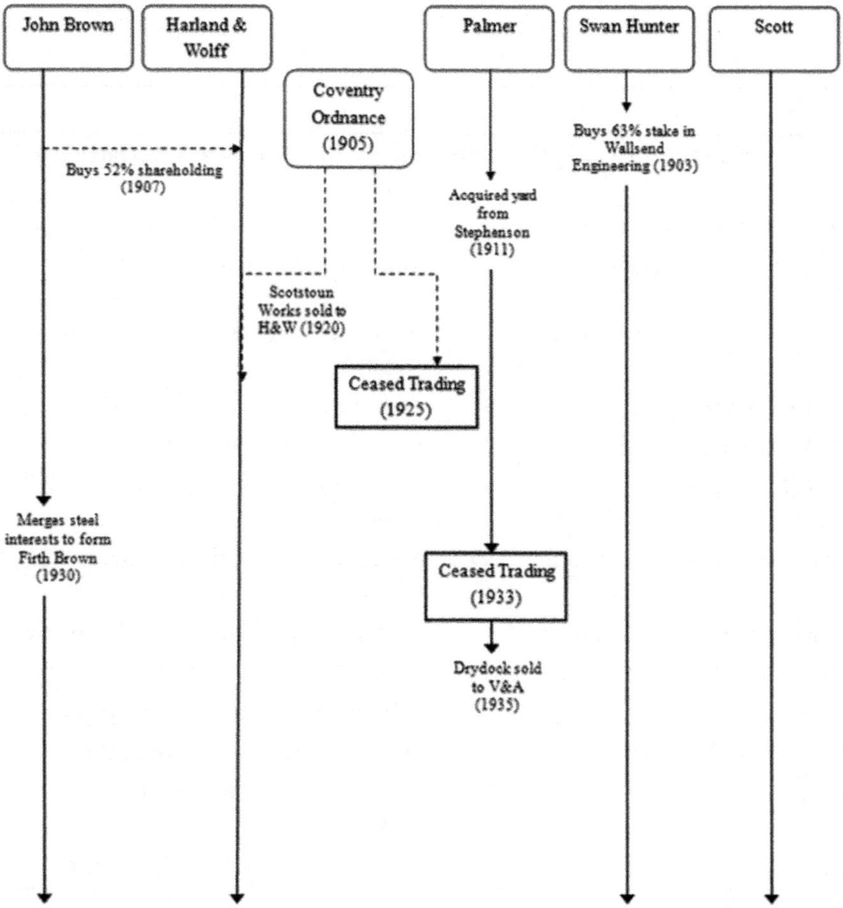
John Brown
Harland & Wolff
Coventry Ordnance (1905)
Palmer
Swan Hunter
Scott
Buys 52% shareholding (1907)
Buys 63% stake in Wallsend Engineering (1903)
Acquired yard from Stephenson (1911)
Scotstoun Works sold to H&W (1920)
Ceased Trading (1925)
Merges steel interests to form Firth Brown (1930)
Ceased Trading (1933)
Drydock sold to V&A (1935)

Appendix 2

Permanent Members of the Supply Board after Manchuria

Name	Position
Air Vice-Marshal Hugh C.T. Dowding (Chairman)	Member for Supply and Research, Air Ministry
Rear-Admiral C.M. Forbes	Third Sea Lord and Controller, Admiralty
Brigadier R.K. Hezlett	War Office & Chairman, Supply Committee I
Air Commodore A.W. Bigsworth	Air Ministry & Chair, Supply Committee V, VI
Mr F.C. Bovenschen	Director of Contracts, War Office
Mr W. Daniels	Controller, India Store Department
Mr C.B. Burdekin	Office of High Commissioner, New Zealand
Lieut-General Sir J. Ronald E. Charles	Master-General of the Ordnance, War Office
Major-General R.L.B. Thompson	War Office & Chair, Supply Committee II
Engineer Rear-Admiral H.A. Brown	Admiralty & Chair, Supply Committee III
Mr J.W.L. Oliver	Admiralty & Chair, Supply Committee IV
Mr W. St. D. Jenkins	Director of Contracts, Admiralty
Mr T. Trumble	Defence Liaison Officer, Commonwealth of Australia.
Mr C.R. Brigstocke	Director of Contracts, Air Ministry

Source: TNA, CAB60/30–1, PSO(SB), 20th–34th meetings.

Appendix 3

DRC Members in 1934

Name	Role and Affiliation
Sir Maurice Hankey	Cabinet Secretary and CID Secretary
Sir Robert Vansittart	Permanent Under-Secretary, Foreign Office
Sir Warren Fisher	Permanent Secretary, Treasury
Admiral Sir Ernle Chatfield	First Sea Lord & Chief of Naval Staff, Admiralty
General Sir Archibald Montgomery-Massingberd	Chief of Imperial General Staff, War Office
Air Chief Marshall Sir Edward Ellington	Chief of Air Staff, Air Ministry

Source: Drawn from Neilson, "Defence Requirements Subcommittee," 652.

Appendix 4

Committee on Defence Policy and Requirements, 1935

Name	Role
Stanley Baldwin MP (Chair)	Prime Minister
J. Ramsay MacDonald MP (chaired July meeting)	Lord President of the Council
Neville Chamberlain MP	Chancellor of the Exchequer
Anthony Eden MP	Foreign Secretary
A. Duff Cooper MP	War Secretary
Viscount Bolton Eyres-Monsell MP	First Lord of the Admiralty
Viscount Swinton MP	Air Secretary
Walter Runciman MP	President of the Board of Trade
Lord Weir	Industrialist
Admiral of the Fleet Sir Ernle Chatfield	First Sea Lord, Chief of Naval Staff (advisory)
Field Marshall A. Montgomery-Massingberd	Chief of the Imperial General Staff (advisory)
Air Chief Marshall Sir Edward Ellington	Chief of the Air Staff (advisory)
Sir Warren Fisher	Permanent Secretary, Treasury (advisory)
Sir Robert Vansittart	Permanent Secretary, Foreign Office (advisory)
Sir Maurice Hankey	CID and Cabinet Secretary (Secretary)

Source: "Committee on Defence Policy and Requirements – Terms of Reference on Defence Requirements," DPR(DR)1, 30 December 1935, CCC: WEIR17.

Bibliography

Unpublished Archival Sources

The National Archives, Kew (TNA)
Admiralty
ADM 1 – Admiralty and Secretariat Papers

Cabinet
CAB 2 – Committee of Imperial Defence, Minutes
CAB 4 – Committee of Imperial Defence, Memoranda
CAB 15 – Committee on the Coordination of Departmental Action
CAB 16 – Ad Hoc Subcommittees of Enquiry
CAB 21 – Cabinet Registered Files
CAB 23 – War Cabinet and Cabinet Meetings
CAB 24 – War Cabinet and Cabinet Memoranda
CAB 27 – Cabinet Committees to 1939
CAB 29 – Naval Conference Ministerial Minutes
CAB 60 – Principal Supply Officers Committee
CAB 102 – War Cabinet and Cabinet: Historical Section (WWII Histories – Civil Series)

Supply
SUPP 3 – Principal Supply Officers Committee (Miscellaneous)

Treasury
T 160 – Finance Files
T 161 – Supply Files

Bank of England – Threadneedle Street, London (BOE)
SMT – Securities Management Trust Papers

Churchill College, University of Cambridge (CCC)
HNKY – Papers of Maurice Hankey
WEIR – Papers of Lord Weir

Cambridge University Library (CUL)
Templewood IX – Papers of Sir Samuel Hoare, First Lord of the Admiralty
Vickers – Papers of Vickers Armstrong

Glasgow University Archive Services – Scottish Business Archive (GUAS)
DC 35 – Sir James Lithgow Papers
DC 96 – Lord Weir Papers
DC 114 – Private Letter Books, Sir Maurice Denny
GD 319 – Papers of Scott's Shipbuilding and Company
GD 320 – Lithgow Investment and Legal Papers
UCS 1 – Papers of John Brown and Company
UGD 4 – Papers of Alexander Stephen and Sons
UGD 100 – Papers of William Beardmore and Company

Mitchell Library, Glasgow City Council (GCC)
UCS 2 – Papers of Fairfield Shipbuilding and Engineering

Basil Liddell Hart Centre for Military Archives, King's College London (BLH)
TB – Papers of Sir Harold Brown and Sir George Turner

Caird Library, National Maritime Museum, Greenwich (NMM)
CHT – Papers of Admiral Alfred Ernle Chatfield
NSS – Papers of the National Shipbuilders' Security scheme
SRNA – Shipbuilders and Repairers National Association (includes Shipbuilding Conference)

Modern Records Centre, Warwick University
MSS 237 – Federated Admiralty Contractors Papers

British Parliamentary Papers

Command Paper 3758 – *Treaty of London of April 22, 1930*, April 1930.
Command Paper 4827 – *Statement Relating to Defence*, March 1935.
Command Paper 4930 – *Exchange of Notes Between His Majesty's Government in the United Kingdom and the German Government Regarding the Limitation of Naval Armament*, June 1935.
Command Paper 5107 – *Statement Relating to Defence*, March 1936
Command Paper 6034 – *Ministry of Supply Bill*, May 1939
Hansard – Series 5, House of Commons

Unpublished Secondary Sources

Davenport-Hines, R.P.T. "The British Armaments Industry during Disarmament, 1918–36." PhD thesis, Cambridge University, 1979.

Murphy, H. "Déjà Vu All Over Again: The Reluctant Rise and Protracted Demise of Scott Lithgow." PhD thesis, University of Westminster, 2001.

Packard, E.F. "Whitehall, Industrial Mobilisation and the Private Manufacture of Armaments: British State-Industry Relations, 1918–1936." PhD thesis, London School of Economics and Political Science, 2009.

Peebles, H. "Warshipbuilding on the Clyde, 1889–1939, A Financial Study." PhD thesis, University of Stirling, 1986.

Published Secondary Sources

Adams, R.J.Q. *Arms and the Wizard: Lloyd George and the Ministry of Munitions, 1915–1916*. London, 1978.

Anderson, D.G. "British Rearmament and the Merchants of Death: The 1935–1936 Royal Commission on the Manufacture of and Trade in Armaments." *Journal of Contemporary History*, 29/1 (1994), 5–37.

Ashworth, W. *Contracts and Finance*. London, 1953.

Baugh, D.A. "Confusion and Constraints: The Navy and British Defence Planning, 1919–1939." In N.A.M. Rodger (ed.), *Naval Power in the Twentieth Century*. London, 1996, 101–119.

Baxter, C.J. "A Question of Blame? Defending Britain's Position in the South China Sea, the Pacific and South-East Asia, 1919–41." *Journal of the Royal United Services Institute*, 142/4 (1997), 66–75.

Bell, C.M. *Churchill and Seapower*. Oxford, 2013.

—. *The Royal Navy: Seapower and Strategy between the Wars*. London, 2000.

—. "Winston Churchill and the Ten-Year Rule." *Journal of Military History*, 74/4 (2010), 74/4 (2010), 1097–1128.

Bell, P. *Chamberlain, Germany and Japan, 1933–4: Redefining British Strategy in an Era of Imperial Decline*. London, 1996.

Berkeley, H. *The Myth that Will Not Die: The Formation of the National Government, 1931*. London, 1978.

Bond, B. *British Military Policy between the Two World Wars*. Oxford, 1980.

Borthwick, A. *Yarrow and Company Limited, 1865–1977*. Glasgow, 1977.

Bowden, S. and Collins, M. "The Bank of England, Industrial Regeneration, and Hire Purchase between the Wars." *Economic History Review*, 45/1 (1992), 120–136.

Boyce, R. *The Great Interwar Crisis and the Collapse of Globalization*. London, 2012.

Boyle, A. *Montagu Norman: A Biography*. London, 1967.

Brandes, S.D. *Warhogs: A History of War Profits in America*. Lexington, KY, 1997.

Brown, D.K. *A Century of Naval Construction: A History of the Royal Corps of Naval Constructors*. London, 1983.

Burton, A. *The Rise and Fall of British Shipbuilding*. London, 1994.
Buxton, I. and Johnston, I. *The Battleship Builders: Constructing and Arming British Capital Ships*. Barnsley, 2013.
Campbell, N.J.M. "Washington's Cherrytrees: The Evolution of the British 1921–1922 Capital Ships." *Warship,* 1/3 (1985).
Cerretano, V. "The Treasury, Britain's Postwar Reconstruction, and the Industrial Intervention of the Bank of England, 1921–9." *Economic History Review,* 62/1 (2009), 80–100.
Chatfield, E. *It Might Happen Again: The Navy and Defence*. London, 1947.
Churchill, W. *The Gathering Storm*. Boston, 1948.
Colledge, J.J. *Ships of the Royal Navy: An Historical Index*. 2 vols. Newton Abbot, 1969.
Coulter, M.W. *The Senate Munitions Inquiry of the 1930s: Beyond the Merchants of Death*. Hartford, CT, 1997.
Dale, I.*Conservative Party General Election Manifestos, 1900–1997*. London, 1999.
Day, L. and McNeil, I. (eds.). *Biographical Dictionary of the History of Technology*. London, 2013.
Dutton, D. "Simon and Eden at the Foreign Office, 1931–1935." *Review of International Studies,* 20/1 (1994), 35–52.
Edgerton, D. *Warfare State: Britain, 1920–70*. Cambridge, 2005.
—. *Britain's War Machine: Weapons, Resources and Experts in the Second World War*. London, 2011.
Elliot, P. *Allied Escort Ships of World War II: A Complete Survey*. London, 1977.
Englebrecht, H.C. and Hanighen, F.C. *The Merchants of Death: A Study of the International Armaments Industry*. New York, 1934.
English, J. *Amazon to Ivanhoe: British Standard Destroyers of the 1930s*. Kendal, 1993.
Fanning, R.W. *Peace and Disarmament: Naval Rivalry and Arms Control, 1922–33*. Lexington, KY, 1995.
Fayle, C.E. *The War and the Shipping Industry*. Oxford, 1937.
Feiling, K.G. *The Life of Neville Chamberlain*. London, 1970.
Ferguson, N. *The Pity of War: Explaining World War I*. London, 1998.
Gibbs, N.H. *Grand Strategy, Volume 1: Rearmament Policy*. London, 1976.
Gordon, G.A.H. *British Sea Power and Procurement between the Wars: A Reappraisal of Rearmament*. Hong Kong, 1988.
Grant, A. *Steel and Ships: A History of John Brown's*. London, 1950.
Greaves, J. *Industrial Reorganisation and Government Policy in Interwar Britain*. London, 2005.
Greenwood, S. "Caligula's Horse' Revisited: Sir Thomas Inskip as Minister for the Co-ordination of Defence, 1936–1939." *Journal of Strategic Studies,* 17/2 (1994), 17–38.
—. "Inskip and Defence, 1936–1939." In Smith, P. (ed.). *Government and Armed Forces in Britain, 1856–1990* (Cambridge, 1996), 155–190.
Hague, A., *Sloops: A History of the 71 Sloops Built in Britain and Australia for the British, Australian and Indian Navies, 1926–1946*. Kendal, 1993.

Hamilton, C.I. "British Naval Policy, Policy-Makers and Financial Control, 1860–1945." *War in History,* 12/4 (2005), 371–395.

Hattendorf, J.B. and Murfett, M.H. (eds.). *The Limitations of Military Power.* London, 1990.

Hayden, T. and Pilisuk, M. "Is There a Military-Industrial Complex Which Prevents Peace? Consensus and Countervailing Power in Pluralistic Systems." *Journal of Social Issues,* 21/3 (1965), 67–117.

Helten, J.V. and Cassis, Y. (eds.). *Capitalism in a Mature Economy: Financial Institutions, Capital Exports and British Industry, 1870–1939.* London, 1990.

Hornby, W. *Factories and Plant.* London, 1958.

Howard, M. *The Continental Commitment: The Dilemma of British Defence Policy in the Era of the Two World Wars.* London, 1972.

Hume, J.R. and Moss, M.S. *Beardmore: The History of a Scottish Industrial Giant.* London, 1979.

Hurtsfield, J. *The Control of Raw Materials.* London, 1953.

Johnman, L. and Murphy, H. *British Shipbuilding and the State since 1918: A Political Economy of Decline.* Exeter, 2002.

— and —. "Welding and the Shipbuilding Industry: A Major Constraint?" In R. Harding (ed.), *The Royal Navy 1930–2000.* London, 2005, 89–116.

— and —. "Subsidy and Treasury: The Trade Facilities Act and the UK Shipbuilding Industry in the 1920s." *Contemporary British History,* 22/1 (2008), 89–110.

Johnson, F.A. *Defence by Committee: The British Committee of Imperial Defence, 1885–1959.* Oxford, 1960.

Jones, G. and Zeitlin, J. (eds.). *Oxford Handbook of Business History.* Oxford, 2008.

Kennedy, P. *The Rise and Fall of British Naval Mastery.* London, 1983.

Koistinen, P.A.C. *Planning War, Pursuing Peace: The Political Economy of American Warfare, 1920–1939.* Lawrence, KS, 1998.

Kyba, P. *Covenants without the Sword: British Public Opinion and Defence Policy, 1931–1935.* Waterloo, ON, 1983.

Landes, D.S. *The Unbound Prometheus: Technological Change and Industrial Development in Western Europe from 1750 to the Present.* Cambridge, 1969.

Ledbetter, J. *Unwarranted Influence: Dwight D. Eisenhower and the Military Industrial Complex.* New Haven, 2011.

Levy, J.P. *Appeasement and Rearmament: Britain, 1936–1939.* London, 2006.

Lindberg, M. and Todd, D. *Anglo-American Shipbuilding in World War II: A Geographical Perspective.* Westport, CT, 2004.

Lloyd George, D. *War Memoirs.* 2 vols. London, 1933.

Lloyd-Jones, R. and Lewis, M. "Armaments Firms, the State Procurement System, and the Naval Industrial Complex in Edwardian Britain." *Essays in Economic and Business History,* 29/1 (2011), 23–40.

Lorenz, E. *Economic Decline in Britain: The Shipbuilding Industry, 1890–1970.* Oxford, 1991.

Lyon, H. "The Relations between the Admiralty and Private Industry in the Development of Warships." In Ranft, B. (ed.). *Technical Change and British Naval Policy, 1860–1939.* London, 1977, 37–64.

Maiolo, J. *Cry Havoc: How the Arms Race Drove the World to War, 1931–1941.* London, 2010.

Marder, A.J. *An Anatomy of British Seapower*. London, 1940.

Marriner, S. "The Ministry of Munitions 1915–1919 and Government Accounting Procedures." *Accounting and Business Research,* 10/1 (1980), 130–142.

Maurer, J.H. "'Winston Has Gone Mad:' Churchill, the British Admiralty and the Rise of Japanese Naval Power." *Journal of Strategic Studies,* 35/4 (2012), 775–796.

McKercher, B.J.C. "The Politics of Naval Arms Limitation in Britain in the 1920s." *Diplomacy and Statecraft,* 4/3 (1993), 35–59.

—. "From Disarmament to Rearmament: British Civil-military Relations and Policy Making, 1933–1934." *Defence Studies,* 1/1 (2001), 24–48.

McMurtie, F.E. (ed.). *Jane's Fighting Ships of World War II.* London, 1997.

McNeill, W.H. *The Pursuit of Power: Technology, Armed Force and Society since A.D. 1000*. Chicago, 1982.

Middlemass, K. *Baldwin: A Biography*. London, 1969.

Miller, C.W. "British naval armaments, cartels, and defence planning between the world wars." *Entreprises et histoire*, 85/1 (2016), 70–87.

Mills, C. Wright. *The Power Elite*. London, 1956.

Mitchell, B.R. *British Historical Statistics*. Cambridge, 1988.

Moretz, J. *The Royal Navy and the Capital Ship in the Interwar Period: An Operational Perspective*. London, 2002.

Morgan, K. *Ramsay MacDonald.* London, 2006.

—. "Cutting the Feet from Underneath Organised Labour? Lord Weir, Mass Production and the Building Trades in the 1920s." *Scottish Labour History,* 43 (2008), 47–68.

Murphy, H. "'No Longer Competitive with Continental Shipbuilders:' British Shipbuilding and International Competition, 1930–1960." *International Journal of Maritime History*, 25/2 (2013), 35–60.

—. "Labour in the British Shipbuilding and Ship Repairing Industries in the Twentieth Century." In R. Varela, H. Murphy and M. van der Linden (eds.). *Shipbuilding and Ship Repairing Workers around the World: Case Studies, 1950–2010* Amsterdam, 2017, 46–116.

Naylor, J. *A Man and an Institution: Sir Maurice Hankey, the Cabinet Secretariat and the Custody of Cabinet Secrecy.* Cambridge, 1984.

Neidpath, J. *The Singapore Naval Base and the Defence of Britain's Eastern Empire, 1919–1941.* Oxford, 1981.

Neilson, K. and Kennedy, G. (eds.). *Far Flung Lines: Studies in Imperial Defence in Honour of Donald Mackenzie Schurman*. London, 1996.

Neilson, K. "The Defence Requirements Subcommittee, British Strategic Foreign Policy, Neville Chamberlain and the Path to Appeasement." *English Historical Review,* 118/477 (2003), 651–684.

Nish, I. (ed.), *Anglo-Japanese Alienation, 1919–1952*. Cambridge, 2010.

Oakley C.A. *The Second City, Glasgow*. Glasgow, 1967.

Overy, R. *The Morbid Age: Britain between the Wars*. London, 2009.

Oxford Dictionary of National Biography. Oxford, 2004.

Parker, R.A.C. "Economics, Rearmament and Foreign Policy." *Journal of Contemporary History,* 10/4 (1975), 637–647.

—. "British Rearmament, 1936–1939: The Treasury, Trade Unions and Skilled Labour." *English Historical Review,* 96/2 (1981), 306–343.

Parkinson, J.R. *The Economics of Shipbuilding in the United Kingdom*. Cambridge, 1960.

Payne, P.L. *Colvilles and the Scottish Steel Industry*. Oxford, 1980.

Peden, G.C. *British Rearmament and the Treasury, 1932–1939*. Edinburgh, 1979.

—. "Sir Warren Fisher and British Rearmament against Germany." *English Historical Review,* 94/1 (1979), 29–47.

—. "The Burden of Imperial Defence and the Continental Commitment Reconsidered." *Historical Journal,* 27/1 (1984), 405–423.

—. "A Matter of Timing: The Economic Background to British Foreign Policy, 1937–1939." *History,* 69/1 (1984), 15–28.

—. *Problems of Setting Strategic Priorities: The Inskip Defence Review of 1937–8*. London, 2010.

—. "Sir Horace Wilson and Appeasement." *Historical Journal,* 53/4 (2010), 983–1014.

Peebles, H. *Warshipbuilding on the Clyde, 1889–1939: Naval Orders and the Prosperity of the Clyde Shipbuilding Industry*. Edinburgh, 1987.

Perry, M. *The Jarrow Crusade: Protest and Legend*. Sunderland, 2005.

Post, G. "Mad Dogs and Englishmen: British Rearmament, Deterrence and Appeasement, 1934–35." *Armed Forces and Society,* 14/3 (1988), 329–357.

Postan, M.M. *British War Production*. London, 1952.

Pugh, P. *The Cost of Seapower: The Influence of Money on Naval Affairs from 1815 to the Present Day*. London, 1986.

Ranft, B. (ed.). *Technical Change and British Naval Policy*. London, 1977.

Reader, W.J. *The Weir Group: A Centenary History*. London, 1971.

Reid, J. *James Lithgow: Master of Work*. London, 1964.

Reid, A. *The Tide of Democracy: Shipyard Workers and Social Relations in Britain, 1870–1950*. Manchester, 2010.

Richardson, D. *The Evolution of British Disarmament Policy in the 1920s*. London, 1989.

Ritchie, S. *Industry and Air Power: The Expansion of British Aircraft Production, 1935–1941*. London, 1997.

Robertson, J.C. "The British General Election of 1935." *Journal of Contemporary History,* 19/1 (1974), 149–164.

Rollings, N. "Whitehall and the Control of Prices and Profits in a Major War, 1919–1939." *Historical Journal,* 44/2 (2001), 517–540.

Roskill, S.W. *Naval Policy between the Wars: The Period of Anglo-American Antagonism, 1919–1929*. 2 vols. London, 1968.

—. *Naval Policy between the Wars: The Period of Reluctant Rearmament, 1930–1939*. London, 1976.

—. *Hankey: Man of Secrets, 1877–1918*. London, 1970.

—. *Hankey: Man of Secrets, 1919–1931*. London, 1972.

—. *Hankey: Man of Secrets, 1931–1963*. London, 1974.

Sayers, R.S. *The Bank of England 1891–1944*. Cambridge, 1976.

Scott, J.D. *Vickers: A History* (London: Weidenfeld and Nicolson, 1962).

Scott, J. "Corporate Control and Corporate Rule: Britain in International Perspective." *British Journal of Sociology*, 41/3 (1990), 351–373.

— and Hughes, M. *The Anatomy of Scottish Capital: Scottish Companies and Scottish Capital, 1900–1979*. London, 1980.

Self, R.C. *Neville Chamberlain: A Biography*. London, 2006.

Shay, R.P. *British Rearmament in the Thirties: Politics and Profits*. Princeton, 1977.

Shizume, M. "The Japanese Economy during the Interwar Period: Instability in the Financial System and the Impact of the World Depression." *Bank of Japan Review*, 2 (2009), 1–10.

Simpson, Michael (ed.). *Documents on Anglo-American Naval Relations, 1919–1939*. London, 2010.

Singleton, J., "Vampires to Skyhawks: Military Aircraft and Frigate Purchases by New Zealand, 1950–70." *Australian Economic History Review*, 42/2 (2002), 183–203.

Sjostrom, W. "Collusion in Ocean Shipping: A Test of Monopoly and Empty Core Models." *Journal of Political Economy*, 97/5 (1989), 1160–1179.

Slaven, A. "A Shipyard in Depression: John Brown's of Clydebank, 1919–1938." *Business History*, 19/2 (1977), 192–217.

—. *British Shipbuilding, 1500–2010: A History*. Lancaster, 2013.

Smart, N. *The National Government, 1931–1940*. London, 1999.

Smith, M.S. "Rearmament and Deterrence in Britain in the 1930s." *Journal of Strategic Studies*, 1/3 (1978), 313–337.

Starkey, D.J. and Murphy, H. (eds.). *Beyond Shipbuilding and Shipping: Britain's Ancillary Maritime Industries in the Twentieth Century*. Hull, 2007.

Sumida, J.T. "British Naval Procurement and Technological Change 1919–1939." In P.P. O'Brien (ed.). *Fighting the Next War at Sea: Technology, Shipbuilding and Future Combat in the Twentieth Century and Beyond*. London, 2001, 128–147.

Tennyson-D'Eyncourt, E.H.W. *A Shipbuilder's Yarn: The Record of a Naval Constructor*. London, 1948.

Thomas, W.A. *The Finance of British Industry, 1918–1976*. New York, 1979.

Thompson, J.A. "The Peace Ballot and the Public." *Albion*, 13/4 (1981), 381–392.

Thorpe, A. *Britain in the 1930s: The Deceptive Decade*. Oxford, 1992.

Todd, D. and Lindberg, M. *Navies and Shipbuilding Industries: The Strained Symbiosis* Greenwich, CT, 1996.

Tooley, T. Hunt. "Merchants of Death Revisited: Armaments, Bankers and the First World War." *Journal of Libertarian Studies*, 19/1 (2005), 37–78.

Toye, R. "The Labour Party and the Economics of Rearmament, 1935–39." *Twentieth Century British History*, 12/3 (2001), 303–326.

Trebilcock, C. "Legends of the British Armaments Industry, 1890–1914." *Journal of Contemporary History*, 5/4 (1970), 3–19.

—. "Science, Technology and the Armaments Industry in the UK and Europe, with Special Reference to the Period 1880–1914." *Journal of European Economic History*, 22/3 (1993), 569–570.

—. *The Vickers Brothers: Armaments and Enterprise, 1854–1914*. London, 1977.

Turner, J.D. (ed.). *Businessmen and Politics: Studies of Business Activity in British Politics, 1900–1945*. London, 1984.

Vickers, R. *The Labour Party and the World: The Evolution of Labour's Foreign Policy, 1900–1951*. Manchester, 2003.

Wark, W. *The Ultimate Enemy: British Intelligence and Nazi Germany*. Ithaca, NY, 1985.

Warren, K.O. *Steel, Ships and Men: Cammell Laird, 1824–1993*. Liverpool, 1998.

Watt, D.C. *Personalities and Policies: Studies in the Formulation of British Foreign Policy in the Twentieth Century*. South Bend, IN, 1965.

Werner, M. *The Military Strength of the Powers*. London, 1938.

Williams, W.H. *Who's Who in Arms*. London, 1935.

Williamson, D.G. "The Ruhr Crisis, 1923–1924." *British Journal of International Studies,* 3/1 (1977), 70–91.

Williamson, P. "'Safety First:' Baldwin, the Conservative Party, and the 1929 General Election." *Historical Journal*, 25/2 (1982), 385–409.

Wilson, B. *The Empire of the Deep: The Rise and Fall of the British Navy*. London, 2000.

Wragg, D. *The Fleet Air Arm Handbook, 1939–1945*. Stroud, 2001.

Wrigley, C.J. (ed.) *A History of British Industrial Relations 1914–1939*. 3 vols. London, 1987.

Index